建筑立场系列丛书 No.72

Workspace Evolution
工作空间进化录

荷兰大都会建筑事务所等 | 编

霍兴花 | 译

大连理工大学出版社

工作空间进化录

004　工作空间在城市中作用的理论综述 _ Andreas Marx

012　蔡国强工作室 _ OMA

022　VSCO奥克兰总部办公室 _ debartolo architects

034　AKQA东京办事处 _ Torafu Architects

042　伍兹·贝格墨尔本工作室 _ Woods Bagot

048　赫罗纳Arquia Banca新办公室 _ Javier de las Heras Solé

056　圣赫罗尼莫街17号办公室 _ CUAC Arquitectura

066　绿色和平北京办公室 _ Livil Architects

076　光速公司总部 _ ACDF Architecture

088　北京海狸工坊办公室 _ MAT Office

100　utopic_US共享办公空间，孔德卡萨尔 _ Izaskun Chinchilla Architects

108　新实验室 _ Macro Sea

114　Rigot Stalars老厂的修复与扩建 _ Coldefy & Associés Architectes Urbanistes

126　3XN新工作室 _ 3XN

140　室内工作环境、办公室和工作室 _ Andrew Tang

150　轻质棚屋，摄影师工作室 _ FT Architects

162　他，她和它 _ Davidson Rafailidis

176　素描、绘画、雕刻小工作室 _ Christian Tonko

186　萨默塞特建筑档案馆和工作室 _ Hugh Strange Architects

194　圣保罗阿克雅工作室 _ Archea Associati

208　建筑师索引

Workspace Evolution

004　Theoretical Review on the Role of Workplaces for Cities_Andreas Marx

012　Cai Guo-Qiang Studio_OMA

022　VSCO Oakland_debartolo architects

034　AKQA Tokyo Office_Torafu Architects

042　Woods Bagot Melbourne Studio_Woods Bagot

048　New Arquia Banca Office in Girona_Javier de las Heras Solé

056　San Jeronimo 17 Office_CUAC Arquitectura

066　Greenpeace Beijing Office_Livil Architects

076　Lightspeed Headquarters_ACDF Architecture

088　Beaver Workshop Office in Beijing_MAT Office

100　utopic_US Co-working Space, Conde de Casal_Izaskun Chinchilla Architects

108　New Lab_Macro Sea

114　Rigot Stalars, Old Mill Rehabilitation and Extension_Coldefy & Associés Architectes Urbanistes

126　3XN's New Studio_3XN

140　Working Interiors, Offices and Studios_Andrew Tang

150　Light Sheds, Photographer's Studio_FT Architects

162　He, She & It_Davidson Rafailidis

176　A Small Studio for Drawing, Painting and Sculpture_Christian Tonko

186　Architecture Archive and Studio, Somerset_Hugh Strange Architects

194　Archea Studio in São Paulo_Archea Associati

208　Index

过去几十年间,创造性工作的特点得到了显著的发展。工作风格的日益灵活通常影响着设计师、艺术家以及专业人士处理日常活动的方式方法。随着移动设备计算能力的不断发展以及几乎无处不在的互联网和云服务,很多工作现在已不再需要固定不变的工作空间,不再有固定的工作时间。这种转变对传统工作空间的功能产生了重要的影响。为了反映工作系统快节奏的变化,办公空间逐渐融入了新的功能和活动内容,包括社交和放松区域,灵活的工作空间以及开放区,在这里,传统的空间层次结构被打破。办公室似乎趋向于跟城市其他公共空间如广场、咖啡店和大堂联系起来。

一方面,办公室通过变得越来越开放、连通、全球化而反映了工作性质结构的变化;另一方面,工作室似乎更专注于空间内的工作活动。工作室是思想产生、发展、经受考验并被赋予生命的创意空间,也是沉思冥想、反复思考和设计创作的地方。工作室中与设计有关的活动通常会涉及社会化场所,如展览馆、陈列室和其他公共活动场所。进行各种任务与活动的现代工作室体现了当今具有创造性和专业性工作的多样性。

工作空间进化录

During the last decades the nature of creative jobs is dramatically evolved. The increasing flexibility of work styles influenced the way designers, artists, and professionals in general approach their daily activities. With growing computational power of mobile devices, and virtually ubiquitous access to internet and cloud services, jobs are being separated from a determined location and time. This shift is having significant consequences for the way traditional workplaces function. To reflect the fast-pace changes of the work systems, office spaces have gradually incorporated new functions and activities, including areas for socialize and relax, flexible working spaces, and open zones where traditional spatial hierarchies are abolished. Offices seem to have a tendency to relate to other urban public spaces, such as squares, coffee shops, and lobbies. If, on the one hand offices are patently trying to reflect a structural change in the nature of work by becoming more open, connected and global, studios, on the other hand, seem to mainly concentrate on the activities carried out in them. Studios are creative spaces where ideas are born, developed and tested, and have been given life. They are also the place where reflection, iterative thinking process, and design production take place. Design-related activities in studios usually expand to become places of socialization, with exhibitions, display and public events. Contemporary studios embody the variety of tasks and activities that characterize the creative and professional work today.

工作空间在城市中作用的理论综述

Theoretical Review on the Role of Workplaces for Cities

"年轻的才子们想要在一个有趣又模糊了工作和社交生活界线的地方工作"（AECOM 2014）[1]

随着新通信技术的出现，人们对生活和工作的看法发生了巨大变化。[2] 互联网和移动技术的日益结合已经改变了地理位置和社会经济活动之间长期以来的明显关系。新的制度结构正在形成，个人日常生活习惯也在打破。此外，我们的消费观念也在急剧变化：从"拥有某物"变成"就想立即使用"。[3] 尽管目前个人的工作业绩不一定非要员工本人出现在办公室才能获得，但设定一个所谓的工作空间的强烈需求仍然存在。此刻，出现经济活动、社会互动、学习活动以及社群形成的各种新的空间概念阐释了我们社会生活的新动态。[4] 过去几十年，我们目睹了工作方式的显著变化，工作空间需要支持这些变化，同时也提出了未来这些空间可能会是什么样的问题。随着新技能和工作方式变得越来越重要，劳动者将扮演什么角色？办公空间以及辅助性基础设施的性质需要如何改变来支持这一变化？以下文章就城市作为工作空间的建筑方法，尤其是由此为世界各地拥有密集的金融和专业服务公司、年轻而技术纯熟的劳动力以及各种各样的工作空间的城市带来的后果进行审视。

谈到工作空间的物理元素，有必要承认这已经成了不确定的概念，因为工作已经开始延伸到各种不同的地点。尽管过去 40 年

"Young talent wants to be in a place that is fun and blurs the boundaries between work and social life" (AECOM 2014)[1]

With the emerge of new communication technologies, the perception of living and working has changed tremendously.[2] The increasing combination of the internet and mobile technologies have altered the long manifested relationship between geographic location and socio-economic activity. New institutional structures are emerging and the everyday routines of the individuals are breaking apart. Furthermore our consumption has accelerated rapidly: the "owning something" moves on to a "just using something immediately".[3] Even though the individual work performance can be obtained without physical presence nowadays, a strong demand for spatial proximity to locate a so called workplace is still present. In this very moment a variety of new spatial concepts in which economic activity, social interactions, learning and community formation emerge, illustrates the new dynamics of our social life.[4] The last few decades have seen significant shifts in work styles and the workplaces needed to support them, raising questions for what these might look like in the future. As new kinds of skills and work styles become more important, what role will the workforce need to play, and how will the nature of office space and the supporting infrastructure need to change to support this? The following issue examines architectural approaches for the city as workplace, in particular the consequences for cities all around the world with their dense concentration of financial and professional services firms, their young and highly skilled workforce, and their diverse range of workspaces.

来知识经济迅猛发展，而研究人员并没有过多关注物理工作空间的新需求。[5] 不过物理空间在过程组织和活动组织中一直是一个关键因素，并最终成为所有组织的有影响力的结构。[6] 在知识经济中，办公环境仍由理性主义范式所统领——工作空间设计简单有效，大多考虑的是让业主和顾客受益而不是天天在这里工作的员工。[7]

过去 30 年，科技已经对我们如何工作产生了越来越深远的影响。1985 年，苹果"Mac"机推出了桌面图标，Windows Excel 发行，第一个".com"域名注册。这些以及其他的科技创新让我们第一次看到了我们的工作将如何改变，但几乎没人能猜到，随着移动电话、笔记本电脑、互联网、电子邮件和社交媒体的随后到来，科技将从根本上改变我们的工作方式。办公室在人们相互交流与合作中所起的作用越来越得到认可，这也反映了从劳动密集型到知识型工作的巨大宏观经济转变。与过去许多工作都是单独完成相比，人们期待办公室能提供更广泛的环境，使大家不论是个人还是团体都能够以更有活力的方式工作。办公室不再是去工作的地方，更多的是去拜访同事并与同事交流的地方。此外，1986 年"大爆炸"的影响意味着伦敦成为许多办公室设计变革的中心。房地产行业采取了新的建筑形式——例如从 1987 年起开始建造的 Broadgate 建筑（1&2 Broadgate 建筑，ARUP）。抬高的楼板、吊顶和大进深平面都是新时代的象征。但这一时期的办公楼设计规格并没有发生显著变化。2008 年的金融危机使人们重新关注成本和对工作空间的使用以

Talking about the physical elements of the workplace, it's necessary to admit that it has become a problematic concept as work began spilling in all sorts of alternative locations. While the knowledge economy has grown tremendously in the last four decades, researchers have not paid much attention to the new needs of the physical workspace.[5] But the physical space remains a critical element in the organization of processes, activities and ultimately the power structure of any organization.[6] In the knowledge economy, office environments are still governed by the paradigm of rationalism – designing workplaces in simple and efficient ways and mostly benefiting the clients or customers rather than the employees occupying it.[7]

Over the last 30 years, technology has had an increasingly profound impact on how we carry out work. In 1985 the Apple "Mac" introduced desktop icons, Windows Excel was launched and the first ".com" was registered. These and other innovations provided the first glimpses of how work was going to change. But few could guess, how fundamentally it would alter work styles – with the later arrival of mobile phones, laptops, the internet, email and social media. Reflecting the wider macro-economic transition from labor intensive work to knowledge-based work, the role of the office was increasingly acknowledged as enabling people to interact and collaborate. The office was expected to provide a wider range of settings in which individuals and groups could work in more dynamic ways compared with much of the more solitary work of the past. The office was becoming less a place to go to work and more somewhere to visit and interact with colleagues. In parallel, the impact of "Big Bang" in 1986 meant that for example London be-

提高企业效率、增加利润和减少损失的需要使人们把重点放在了房地产的实际成本上，而且许多组织力图让大家缩减空间需求。移动技术的进步意味着在办公室之外的移动工作和远程工作不再受到不可靠的技术或糟糕的连接的阻碍。

当今，对劳动力需求的日益增加可以被看作全球化的结果和技术对工作活动的影响，这改变了劳动力的现状。员工通常更年轻，资质也更高。关于 X 一代和 Y 一代对工作空间的影响，很多人都写过文章，但是在这场辩论中，有些关于一代人的神话，忽视了更广泛的社会变革所带来的影响。尽管现在的工作空间中有四代人，但他们的期望是一样的。为他们提供选择性和灵活性成为吸引最优秀人才的首要任务之一。有四个方面最为重要：

· 选择工作空间，不论是办公室内部还是外部，无处不在的无线连接促进了这一选择
· 分享知识、合作和社交的机会
· 到达工作空间高效、廉价的交通设施
· 周边良好的便利设施，包括咖啡店、零售店和餐馆

came an epicenter of many of the changes in office design. The property industry responded with a new built form – for example the Broadgate buildings of 1987 onwards (1&2 Broadgate by ARUP). Raised floors, drop ceilings and deep plan floors were all symptoms of the new era. The office building specification has not changed significantly since this period. The financial crisis in 2008 created a renewed focus on cost and the use of workplace to improve corporate efficiency. The need to address profit and loss pressures led the focus on the real cost of property, and many organizations pushed for notable reductions in space demand. Advances in mobile technology meant that working on the move and working remotely, outside the office, was no longer hindered by unreliable technology or poor connectivity.

Nowadays, the increasing demands of the workforce can be seen as result that followed in terms of globalization and the impact of technology on work activity which transformed the profile of the workforce. Employees are generally younger and more highly qualified. Much has been written about the impact of Generation X and Generation Y on the workplace, but there is something of a generation myth in this debate which ignores the impact of wider social changes. While we now have four generations in the workplace, they are quite alike in expectations. Providing them with choice and flexibility became one of the major priorities to attract the best talents. There are four aspects that count the most:

- Choosing the place to work – inside and outside the office – which is facilitated by ubiquitous connectivity

1&2 Broadgate建筑,ARUP,1987年
1&2 Broadgate by ARUP, 1987

・进入休闲设施如酒吧、剧院、健身房的通道

最根本的环节是"选择"。当今的劳动者在各个层面上都要求有选择权,包括职业发展、当地零售店、工作生活的平衡以及休闲娱乐等等,这些对选择的期望越来越多地反映在工作空间上,雇主们希望在其中提供更多的服务。工作空间的形状也朝着更"积极的设计"改变,以确保人们在工作中可以更多地走动。此外,越来越多的为特定项目和专门技术或短期内补充公司内部资源而雇用的合同工或自由职业者也对工作空间产生了影响。由于公司边界越来越不明显,工作空间的质量、位置和周边设施也必须能够吸引这些员工。

恰在此时,共享办公的出现开始改变办公环境。共享办公是近年来迅速发展起来的一个广义的术语。[8] 该术语指的是大家在灵活共享的办公环境中"并肩"工作的做法。在共享办公中,办公桌可以根据不同情况进行租用,在这里,志同道合者形成团队。作为一种城市现象,共享办公空间主要是在城市中发展起来的,它鼓励合作、创新、想法共享、指导、交流、社交,并为小公司、创业公司以及普遍缺少大型组织机构资源的自由职业者创造新的商业机会。[9] 汇总目前学术界有关共享办公的概念和评论,我们得出如下有关办公空间的定义:

"共享办公是关注企业家、创新型工作者和知识型工作者需求的、灵活的商业和工作模式。任何共享办公空间都是基于合作、团

- Opportunities for sharing knowledge, collaborating and socializing
- Efficient and inexpensive transportation supplies to the workplace
- Good local amenities including coffee shops, retail and restaurants
- Additional access to leisure facilities such as bars, theatres, gyms

The fundamental link is "choice". The workforce nowadays demands choice at all levels: career development, local retail, work-life balance, leisure and so on. This expectation of choice is increasingly reflected in the workplace, where employers are seeking to provide more services. The shape of the workplace is also changing towards a more "active design" to ensure that people move more at work. Furthermore, the increasing number of contract or freelance workers employed for specific projects and expertise, or to supplement in-house resources in the short-term will also have an impact on the workplace. As corporate boundaries are becoming more permeable, the quality of the workplace, location and the surrounding facilities have to attract these workers too.

In this very moment, the emerge of coworking space is changing the office environment. Coworking is a broad term that has been rapidly expanding in recent years.[8] The term refers to the practice of working "alongside each other" in a flexible and shared office environment where desks can be rented on a different basis and where like-minded professionals form a community. As being an urban phenomenon, Coworking spaces have been developed mainly in cities to encourage collaboration, creativity, idea sharing, mentoring, networking, socializing and generating new

1. AECOM (2014): See Further: the Next Generation Occupier
2. Virilio, P. (1995): La vitesse de libération. Essai. Paris: Galilée.
3. Rosa, H. (2010): Alienation and Acceleration. Towards a Critical Theory of Late-Modern Temporality. Malmö/Arhus: NSU Press.
4. Waters-Lynch, J. et al. (2016): Coworking: a transdisciplinary overview. RMIT University research.
5. Ashkanasy, N, Ayoko, O, & Jehn, K (2014): 'Understanding the physical environment of work and employee behavior: An affective events perspective', Journal of Organizational Behavior, vol. 35, no. 8, pp. 1169–1184.
6. Foucault, M (1977): Discipline and punish: The birth of the prison. Vintage.
7. Knight, C. & Haslam, SA. (2010): 'Your place or mine? Organizational identification and comfort as mediators of relationships between the managerial control of workspace and employees' satisfaction and wellbeing', British Journal of Management, vol. 21, no. 3, pp. 717–735.

体、可持续性、开放性及可及性这五个价值观。除了具体的工作空间，建立交流、创新和教育网络，为新企业家提供支持处于最显著的位置。"10

关于共享办公的这一解释只是一个例子，说明目前工作空间的设计和管理如何通过提供独特的体验和便利设施以及持续的适应性来应对工作的选择性和多样性。目前的办公室需要始终如一地集中精力解决以下问题：

·从固定、长期租赁的空间到灵活、随需应变的空间的转变
·更少空间，使用更高效，更有效
·空间成为表达企业文化和价值的媒介
·为持续的适应性和多样化的使用模式而设计
·基于活动的工作空间为员工提供合作、专心工作、交流、创新、保密工作和沉思的空间
·共享空间的使用是促进合作的手段

business opportunities for small firms, start-up companies and freelancers who typically lack the resources of large organizations.[9] Summarizing several concepts and comments about Coworking Spaces in the present academic discours, the following definition tackles the main aspects:
"Coworking is an integrative and flexible business and work model which focuses on the demands of entrepreneurs, creative and knowledge workers. Every Coworking Space underlies the five values collaboration, community, sustainability, openness and accessibility. Alongside of the concrete workplace, the setup of networks for exchange, innovation and education stands in the foreground which supports the new entrepreneurs".[10]
This explanation on coworking is just one example to show, how the design and management of workplaces nowadays must respond to choice and diversity by providing continuous adaptability, as well as delivering uniqueness of experience and amenities. The present office needs are focused consistently on addressing the following issues:
- A shift from fixed, long-term leased space to flexible and on-demand space
- Less space, used more efficiently, and more effectively
- Space being a medium for expressing corporate culture and values
- Design for continuous adaptability and diverse usage patterns
- Activity-based workspaces providing for collaboration, concentration, communication, creativity, confidentiality and contemplation

8. Botsman, R., & Rogers, R. (2011): What's mine is yours. New York, NY: Collins. / Ferriss, T. (2009): The 4-hour workweek, expanded and updated. New York, NY: Crown.
9. Spinuzzi, C. (2012): Working alone together coworking as emergent collaborative activity. Journal of Business and Technical Communication, 26(4), 399-441.
10. Schürmann, M. (2013): Coworking Space - Geschäftsmodell für Entrepreneure und Wissensarbeiter. Springer Gabler Verlag, Wiesbaden.
11. De Croon, E., Sluiter, J., Kuijer, P. P., & Frings-Dresen, M. (2005). The effect of office concepts on worker health and performance: a systematic review of the literature. Ergonomics, 48(2), 119e134
12. Leaman, A., & Bordass, B. (1999). Productivity in buildings: the 'killer' variables. Building Research & Information, 27(1), 4e19
13. Seddigh, A./ Stenfors, C./ Berntsson, E./ Baath, R./ Sikström, S./ Westerlund, H. (2015): The association between office design and performance on demanding cognitive tasks. Journal of Environmental Psychology. Elsevier Ltd.

·提供便利设施和服务
·创造和管理难忘的体验以吸引人才

办公室设计的选择取决于经济激励措施和绩效概念。尽管研究显示了不同的结果[11],但人们普遍认为与窄小的隔间办公室相比,开放式平面布局的办公室更能促进沟通交流。虽然研究表明办公室设计可以提高员工绩效的15%[12],但白领的绩效测定通常比工作空间成本更难评估。这妨碍了对于拆除墙体实施开放式办公设计的经济结果的评估。[13] 在这种背景下,一个重要教训是员工人数增长和空间需求的传统关系正在改变,导致"无限增长"。这一问题目前必须与创造新的工作机会一起得到解决。这将使组织机构既能吸引人才又能更好地管理他们的房产,提高办公空间的灵活性和适应性。正如一些项目将要展示的那样,如蔡国强工作室和圣赫罗尼莫街17号办公室,对工作空间的理解是多样化的。本书将探讨当今办公空间是如何设计的、作为创新互动区域是如何发挥作用的以及如何将劳动者和城市联系在一起。

- Use of shared spaces as a means to facilitate collaboration
- Provision of amenities and services
- Creating and managing memorable experiences to attract talent

The choice of office design depends on both economic incentives and ideas about performance. Although research shows mixed results[11], it is a widespread idea that communication improves in open-plan offices in comparison to cell offices. Although it has been suggested that the office design may explain up to 15% of employees' performance[12], measures of performance for white-collar workers are usually more difficult to assess than workspace costs. This impedes estimation of the financial consequences of removing the walls and implementing an open-plan office design.[13] Within this context, a key lesson is that the traditional relationship between headcount growth and space demand is changing, resulting in "spaceless growth". This issue must be approached in the present – together with the establishment of new opportunities to work. This will enable organizations to both attract talent and manage their real estate with increased flexibility and adaptability. As some projects will show – such as the Cai Guo-Qiang Studio or the San Jeronimo 17 Office – the understanding of workplace is very heterogeneous. This book will examine how office spaces are designed nowadays, how they function as zones for creative interaction and how they serve as link between the workforce and the city. Andreas Marx

位于纽约的蔡国强工作室由OMA合伙人重松象平负责进行了改造和扩建。改造后的工作室围绕阳光充足的中央庭院组织规划其各种功能。工作室位于纽约下东区，改建自始建于1885年的一所校舍，原先只有地面一层。扩建后的空间包括地下室和位于第一大街上的相邻店面，占地面积翻了一番。扩建后的工作室增加了艺术家创作、展览和接待的空间。

首席设计师重松象平说："一系列独特的垂直交通流线空间将光线从街道层引入了地下空间。自然光通过由可上人的玻璃板和可反射光线的竹林组成的采光井照入地下。在建筑内部，一部中央楼梯通往下面楼层和一个两层高的展示空间。一个原有的结构性拱顶被改造为'潜望镜'，通过镜面折射建立了图书馆和街道间的视觉联系。"

对于没有画廊代理的艺术家来说，这个806m²的工作室兼具创作、存档、展览、接待、管理和办公功能，在其业务运作中起着至关重要的作用。不同空间的功能本就相互渗透，而相同的建筑材料和连续的光线淡化了它们之间的界线，将整个工作室连为一体。树脂墙沿着庭院一侧上下贯通，成为分散自然光线的中心柱体。内置灯具和机械设备结合在墙体内，创造出储存、展示和工作空间，沿墙体长轴方向的不同位置还设置了多个支持空间。这次改建还保留了许多历史建筑元素，包括红色的校门（现仍作为主要入口）、原始的砖石建筑、拱门以及楼梯和铁栏杆。每个房间功能灵活，既可以作为私人空间，也可以当作公共场所。拥有博物馆照明质量的两个画室用于进行日常展示以及作为提供餐饮的接待区。地下空间有一个会议室，配备了音响系统、投影设备和5m长的花旗松实木桌子，可以举行大型会议和电影放映会。图书馆是接受出版商和记者采访的专属空间，这里收集了蔡国强所有的出版作品。用日本经典元素榻榻米和浮竹苇灯打造的茶室，既是主人与访客礼仪性交往的传统空间，也是给所有工作人员使用的现代冥想空间。

Cai Guo-Qiang Studio

The renovation and expansion of Cai Guo-Qiang's studio in New York, led by OMA Partner Shohei Shigematsu, organizes the studio's multiple functions around a central, light-filled courtyard. The studio, located in a converted 1885 schoolhouse in New York's Lower East Side, previously consisted of one floor on the ground level. The expanded space now includes both the basement level and an adjacent storefront on First Street. Now twice the original square footage, the expanded headquarters enhances the artist's capability for production, exhibition and reception.

"A series of distinct vertical connections from the street level illuminate the cellar level below," commented lead designer, Shohei Shigematsu. "Natural light from the courtyard filters down through a series of light wells composed of walkable glass panels and a reflective bamboo vault. Inside, a central stairwell provides circulation to the lower level, as well as a

蔡国强工作室
OMA

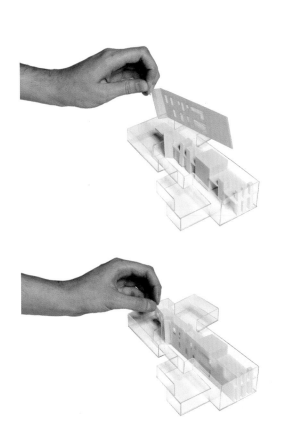

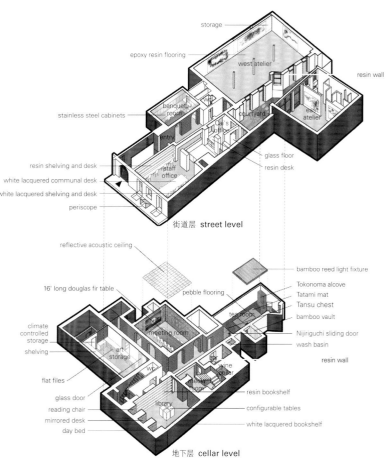

double-height display space. An existing structural vault was repurposed as a periscope, offering views to the street via a mirrored desk in the library."

Without gallery representation, the 806-square-meter studio plays a crucial role in the artist's operations, acting as the main work, archive, gallery, reception, administration, and office space. The porous boundaries between the programs unite the studio through continuous materials and light. The resin wall spans both levels along the courtyard edge, acting as a central spine distributing natural light. The wall system is constructed with integrated lighting and mechanical infrastructure, incorporating storage, display, workspace, and support spaces at various points along its length. The renovation also preserves many historic elements of the building,

including the red school door which continues to serve as the main entry, original brick and stone masonry and archways, and the existing stairs and iron railings. Each room has the flexibility to function as both a private workspace and public venue. Two ateliers with museum quality lighting are used for daily exhibitions as well as catered receptions. In the cellar, a new space with A/V capabilities and a 5-meter solid Douglas Fir table hosts large meetings and film screenings. The library provides a dedicated space for interviews with publishers and journalists with Cai's collection of publications close at hand. A tea room composed of the essential elements of a customary Japanese tea house. Tatami mats and a floating bamboo reed light fixture.acts as a traditional ceremonial space for visitors as well as a modern contemplation space for the studio.

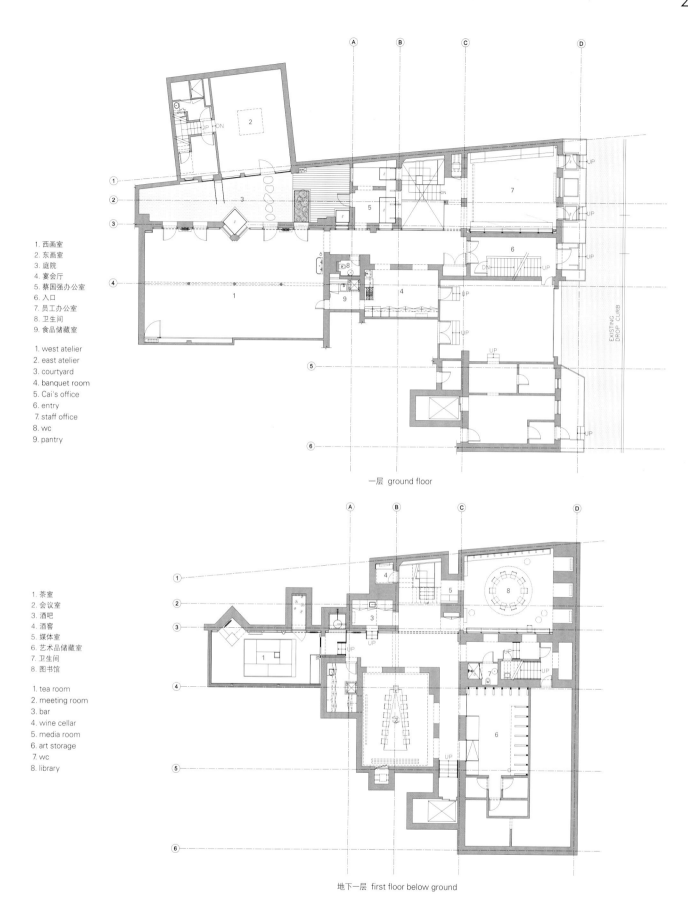

1. 西画室
2. 东画室
3. 庭院
4. 宴会厅
5. 蔡国强办公室
6. 入口
7. 员工办公室
8. 卫生间
9. 食品储藏室

1. west atelier
2. east atelier
3. courtyard
4. banquet room
5. Cai's office
6. entry
7. staff office
8. wc
9. pantry

一层 ground floor

1. 茶室
2. 会议室
3. 酒吧
4. 酒窖
5. 媒体室
6. 艺术品储藏室
7. 卫生间
8. 图书馆

1. tea room
2. meeting room
3. bar
4. wine cellar
5. media room
6. art storage
7. wc
8. library

地下一层 first floor below ground

项目名称：Cai Guo-Qiang Studio / 地点：New York, USA / 首席建筑师：Scott Abrahams / 合伙人负责人：Shohei Shigematsu / 项目团队：Ted Lin, Lawrence Siu, Ian Mills, Matthew Austin, Hanying Zhang, Nick Demers-Stoddart, Cass Nakashima, Sean Billy Kizy / 执行建筑师：Shiming Tam Architect PC / 结构工程师：Robert Silman Associates PC / 机电管道工程师：Plus Group Consulting Engineering PLLC / 照明顾问：Tillotson Design Associates / Dot Dash / 承包商：P&P Interior Inc. / 建筑面积：3,831m² / 有效楼层面积：573m² / 竣工时间：2015

摄影师：©Brett Beyer (courtesy of the architect) (except as noted)

VSCO奥克兰总部办公室
debartolo architects

由年轻的摄影师和创意人员组成的团队在加州奥克兰创建的VSCO (Visual Supply Co.) 公司迅速发展成为摄影行业最大的品牌和增速最快的公司 (FastCompany 2014)。公司的理念就是在自然的环境中展现企业的文化与目标，这成为此次设计的首要驱动力。业主、建筑设计师和建造者强强联合，创造了独具匠心、无与伦比的创新工作区。设计遇到的真正的挑战是在创业者的家乡奥克兰找到一个能安置这个创意巨头的地方，最后被列入国家历史遗迹名录的两栋大楼中的一栋入选。

位于奥克兰市中心百老汇1500号的建于1922年的鲁斯兄弟服装店是一座哥特复兴式建筑，由威廉·诺尔斯设计，因其上半部分用粉色

上釉陶土装饰而与众不同。该建筑最初被设计为一家百货商店，鲁斯兄弟服装店占据一层主要的角落和全部上层空间，另外两家小商店占据建筑最北端临百老汇大街的隔开的一层空间。

找到地方后，接下来就是为打造VSCO新总部做出努力：精简杂乱的楼内结构，保留其原始空间的精华，使其回归原有的结构和VSCO新总部的设计精神。在进行空间规划，回应结构性网格与日光模式的关系问题时，建筑师遇到了原有不透明混凝土抗震墙带来的挑战。

仔细去粗取精之后，建筑师只增加了基本的要素以满足作为二、三层租户的VSCO公司的直接需要——公司运转及办公功能。

debartolo建筑师事务所提出了几个推动设计的有启发性的想法：

· 精心设计一个振撼的入口体验，瞬间将来访者与VSCO公司的企业精神连在一起。

· 优化边缘空间，将建筑物的生命活力带到大街上，并最大化地利用动态的日光质量。

· 将楼内的主要工作空间与供他人免费使用的外部空间隔离。

公司的一个口号是"让我们共创美好"，而公司的空间形象证明了他们对美好工作环境的愿望带来的成功。重新利用的花旗松木厚地板，热轧钢拱腹和橱柜，精心布局的管道系统，贴心的向上照明设计，精挑细选的家具，使整个空间流畅飘逸，为八十多名员工提供了高效的工作空间。

VSCO Oakland

Visual Supply Co. (known as VSCO) founded by a youthful team of photographers and creatives in Oakland, California, has quickly become one of the largest brands and fastest growing companies in the photographic industry (FastCompany 2014). Its ethos of manifesting culture and purpose in the physical environment became the primary driver of the design, and combined with a strong collaborative relationship between owner, architect, and builder, yielded a uniquely innovative workspace. The real challenge was finding a space to house this creative giant in the founder's home city of Oakland, however one of two buildings on the National Registry of Historic Places became available.

The 1922 Roos Brothers Clothing Store, located at 1500 Broadway in downtown Oakland, is a Gothic Revival building designed by William Knowles, unique for its glazed pink terra cotta ornament on the upper stories. The building was designed originally as a department store with Roos Brothers occupying the main corner ground floor space and all upper floors, and two smaller stores occupying separate ground floor spaces at the north end on Broadway.

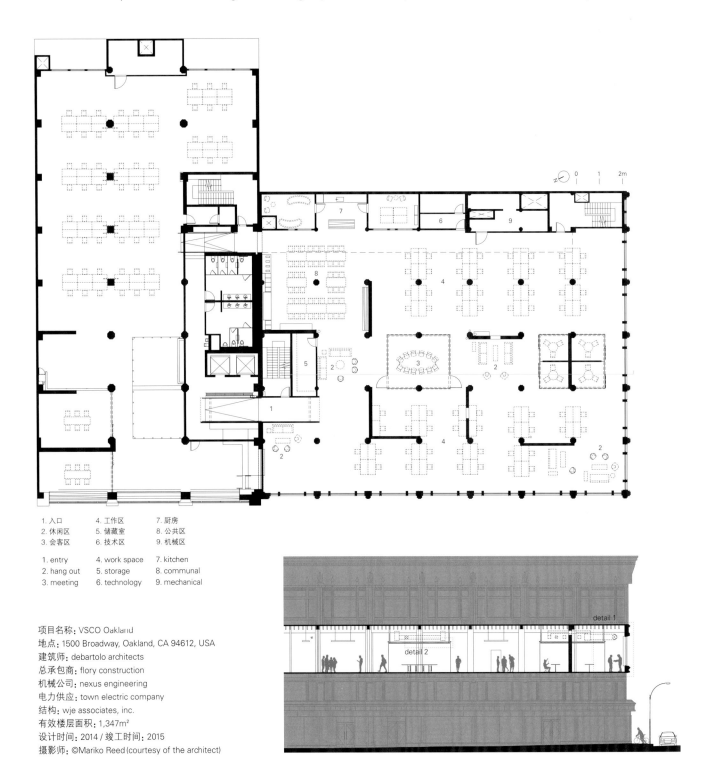

1. 入口　　4. 工作区　　7. 厨房
2. 休闲区　5. 储藏室　　8. 公共区
3. 会客区　6. 技术区　　9. 机械区

1. entry　　4. work space　7. kitchen
2. hang out　5. storage　　8. communal
3. meeting　6. technology　9. mechanical

项目名称：VSCO Oakland
地点：1500 Broadway, Oakland, CA 94612, USA
建筑师：debartolo architects
总承包商：flory construction
机械公司：nexus engineering
电力供应：town electric company
结构：wje associates, inc.
有效楼层面积：1,347m²
设计时间：2014 / 竣工时间：2015
摄影师：©Mariko Reed (courtesy of the architect)

Once found, the process of stripping back existing interior clutter began an effort to distill the raw space back to its true structure and spirit for the new VSCO headquarters. The existing opaque concrete seismic walls challenge the architects in how space planning might respond to structural grid in relationship to daylight patterns.

Carefully distilled, only essential elements were added back to support the direct needs, flow and function of VSCO as a tenant who now occupies the second and third floors in.

The studio of debartolo architects came up with several driving ideas that propelled the design:

- Craft a strong entry experience that immediately connects the visitor to the VSCO ethos.

- Optimize the spaces along the edge to bring the life of the building to the street and maximize the use of dynamic daylight qualities.

- Insulate the major workspace at the interior of the building from the exterior with other complimentary uses.

One of the slogans of the company is, "Let's build something beautiful together", and the images of the space speak to the success of their aspirations for their work environment. Through re-purposed Douglas Fir wood plank floors, hot rolled steel soffits and cabinets, carefully orchestrated conduit runs and ductwork, thoughtful uplighting, and carefully selected furniture pieces – the space flows smoothly and fluidly as an efficient workspace for over 80 people.

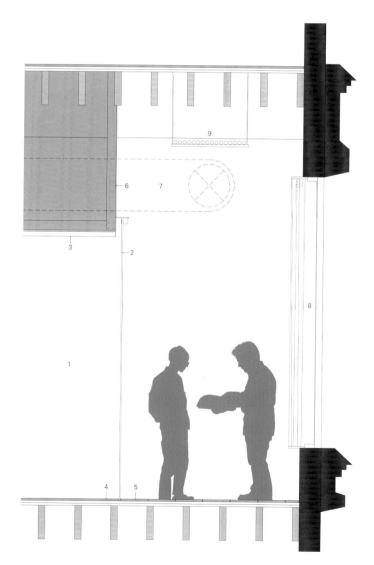

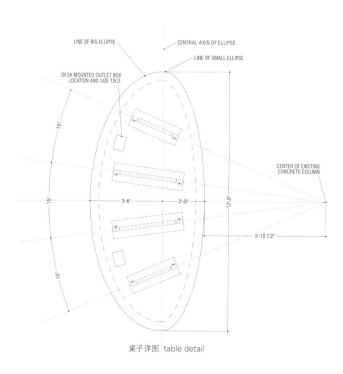

桌子详图 table detail

1. conference room 2. standard glass partition mounted in custom steel frame 3. custom MDF acoustic ceiling 4. carpet tile 5. reclaimed FIR plank flooring 6. steel soffit facing with custom perforated MECH. grilles 7. exposed mechanical 8. plywood window "buckets" 9. data / power raceway

详图1 detail 1

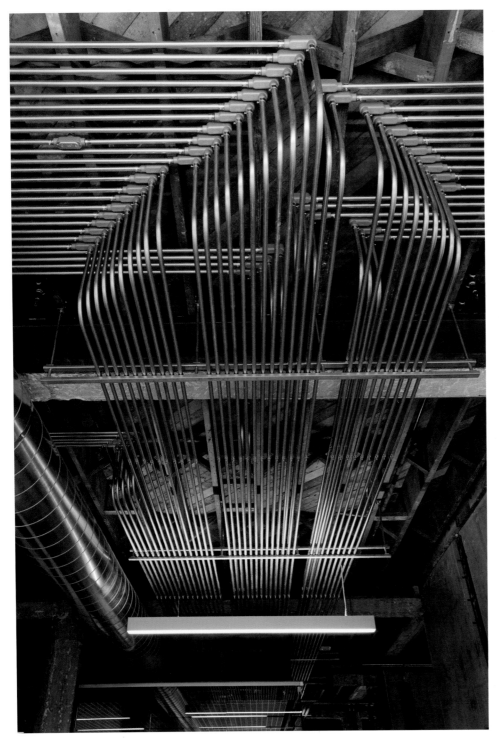

1. 5/8" plywood blocking painted black
2. provide mill finished steel angle 3"x5"x1/4"
3. typical lighting and HVAC fixture as occurs. see RCP.
4. custom steel angle beyond
5. provide blocking as required. to support routing of duct and utilities to fixtures. refer to RCP.
6. coordinate framing of soffit ceiling with architect in field based on equipment requirements in soffit. suspended system or typ. fixed framing. provide acoustic insulation above ceiling
7. custom CNC routed perforated acoustic ceiling system. pattern similar to obersound 5.5 "rain" acoustic panel.
8. 16 GA. hot rolled steel panel
9. typ. #12 pan head phillips screw, black-oxide 18-8 stainless steel – layout equally spaced on all sides
10. pre-fabricated 1/4'' mill finished steel plate angle
11. typical dorma frameless head jamb. refer to A9 series
12. perforated plywood system, as occurs. refer to door finish schedule. coordinate with architect in field

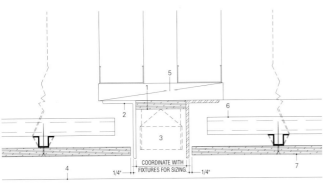
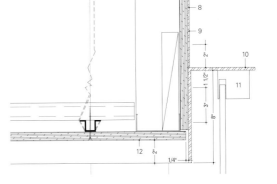

详图2 detail 2

AKQA东京办事处
Torafu Architects

Torafu 建筑师事务所为 AKQA 这家有全球视野和富有创新精神的公司的东京办事处做了室内空间设计。因为办公室位于负一层,建筑师力图创造一个有吸引力的环境,使员工可以在这个拥有开放的空间、6m高的天花板和自然采光的办公室内追求发展和进行创造性的活动。

建筑师利用原有的平整地形,以一个种满大树的内院为中心,创造了一个拥有不同分区的办公空间,每个分区都有着独特的功能。访客走进办事处,首先映入眼帘的是由 Torafu 建筑师事务所设计的"水气球"灯,灯悬挂在接待区可以俯瞰整个办公空间的平台上。与接待区在同一高度还有一个会客厅,而拥有钢框门的三个私人会议室则被设在下面楼梯口处。

工作区被设置在场地内一个比较紧凑的区域内,大部分空间由两个长桌占据,长桌挨近内院且相对设置。这种设计能为办公空间提供足够的光线,员工开会时可以清楚地看清彼此。当举行会议时,这种环形的流线设计会让内院周围的空间有浑然一体之感。此外,位于白色角落的一个覆盖瓷砖的酒吧柜台和一家摄影工作室为办公室赋予了更多的功能。

磨光的混凝土地板、从内院照射进来经过砾石反射的自然光、施工脚手架地面制成的大楼梯和舞台一样的工作空间、彩色玻璃和放在沙袋里由白色砂浆固定的植物等,这些来自不同区域的设计元素互为补充,使整个空间浑然天成。另外,建筑师在可移动的塔式脚手架上安装了篮球筐,建造出来一个娱乐区。

建筑师充分利用光线充足的内院及原有平整地形的开阔空间,同时谨慎利用每个空间而不显得太满,目标是打造一个彰显该公司追求发展和创造性的空间。

AKQA Tokyo Office

We designed the interior space for the Tokyo office of AKQA, a global ideas and innovation company. Since the office is located on the first basement floor, we strived to create an appealing environment where the agency could pursue its developmental and creative activities by using the site's open spaces, six-meter high ceilings and natural lighting. Centered around an inner court with big trees, we sought to capitalize on the existing leveled topography to make it work as an office by creating distinct zones with distinct functionalities. Visitors are greeted by "water balloon" lights, designed by Torafu, which can be found suspended throughout the reception area on the raised platform overlooking the entire office space. On the same level is one room for meeting

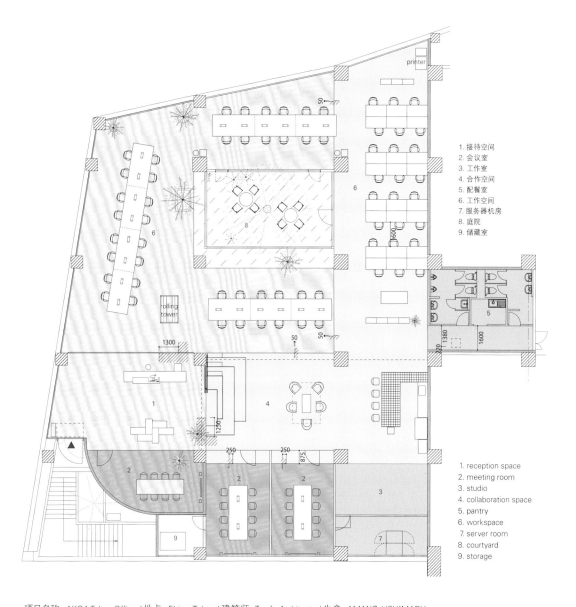

1. 接待空间
2. 会议室
3. 工作室
4. 合作空间
5. 配餐室
6. 工作空间
7. 服务器机房
8. 庭院
9. 储藏室

1. reception space
2. meeting room
3. studio
4. collaboration space
5. pantry
6. workspace
7. server room
8. courtyard
9. storage

项目名称：AKQA Tokyo Office / 地点：Ebisu, Tokyo / 建筑师：Torafu Architects / 生产：AMANO / ISHIMARU
灯具：Izumi Okayasu / 照明设计：E&Y, gleam / 植物：SOLSO / 用途：office
有效楼层面积：710m² / 设计时间：2014.08—2014.11 / 施工时间：2014.11—2015.1 / 摄影师：©Daici Ano (courtesy of the architect)

visiting clients, while three private meeting rooms with doors framed by steel sash windows can be found at the bottom of the grand stairs.

While workstations are found in a compact area of the site, most of the space is taken up by a collaboration space consisting of two long tables adjacent to the inner court, which provides enough light for coworkers to clearly see each other when holding meetings. This circling line of flow brings a sense of unity to each space found around the court. Moreover, a tile-clad bar counter and a photo shoot studio set up in a corner painted in white opens the office to a variety of uses.

Elements from each area, such as polished concrete floors, white gravel diffusing natural light from the inner court, a grand staircase and a stage-like work space made from construction scaffolding flooring, the accent colored glass, and planters made of white mortar casted in sandbags, etc., come together to make the space whole while complementing

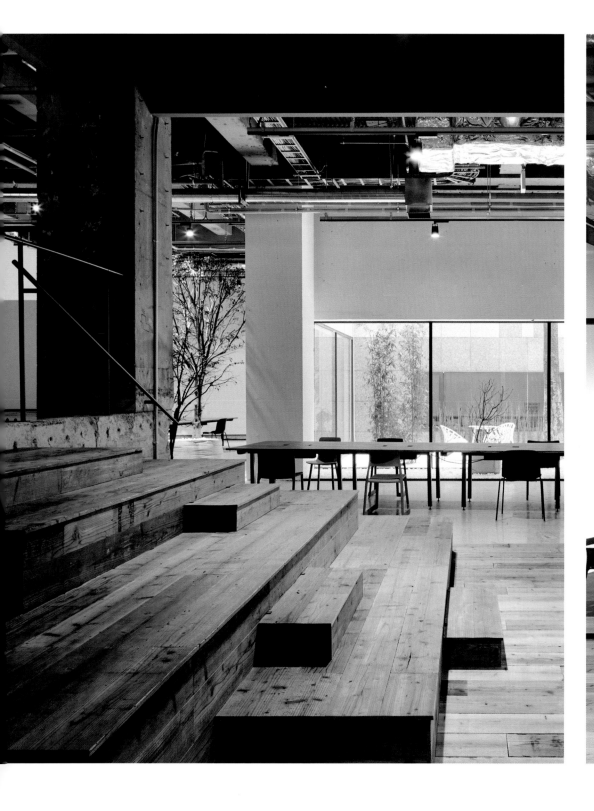

each other. Furthermore, we sought to contribute to in-office entertainment by installing a basketball hoop on a rolling tower scaffold.

By turning the light-filled inner court and the leveled topography of the existing open space site into its greatest appeal factors, and performing careful additions to each area so as not to appear too finished, our aim was to create a space hinting at the agency's developmental and creative activities.

Torafu Architects

伍兹·贝格墨尔本工作室
Woods Bagot

有比设计自己的办公室更费劲的项目吗? 让一群设计理念迥异的建筑师和室内设计师最终设计出完美的作品是一件非常具有挑战性的工作。

为了避免出现这种情况,伍兹·贝格墨尔本工作室采用了一个协商式工作空间战略,即向所有在该环境中工作的员工收集反馈信息。反馈信息表明,员工的首要愿望是既能协同工作也能社交聚会。虽然这对装修提出了规划要求,但仍然缺少项目概念的远景。

正是考虑到这一点,设计团队想到了由Ferran Adria所著的那本简单却又充满灵感的烹饪书《家庭餐》。受每天在传说中的El Bulli餐厅所吃的饭菜启发,这本书集中研究了一天中最重要的时刻——大厨们聚在一起交流、分享和协作。

伍兹·贝格的负责人与设计总监Bruno Mendes说,该项目体现的是家庭情怀和亲密感。"大家聚在一起来分享体验,这一简单的概念成为我们工作室的中心理念。我们要做的是发挥和强化工作室的集体协作精神。"

基于这种团聚的概念,工作室内的各种正式和非正式空间都围绕着楼板中心的一大块中心区域建造,该区域将各种正式和非正式的工作区连在一起,形成了一个聚集空间,通过连接工作室两个楼层的方式起着开放礼堂的作用;它也为受邀嘉宾发言、员工看电影及周五晚上举办舞会提供了平台。员工主餐厅区与工作区的连通对于加强社交和合作至关重要。

在材料的选择上,无论在造型能力方面还是在应用方面,木材都是最佳选择。在定义工作室每个关键空间方面,木材和作为过渡元素的钢板交替使用来定义各个空间高度的变化,其直观结果就是把工作室生活与工作的细微差别表现出来。

Woods Bagot Melbourne Studio

Is there a project that is more demanding than designing your own office? Having a group of architects and interior designers pulling in different directions for design supremacy was always going to be a challenge.

To avoid such a situation, Woods Bagot Melbourne studio embarked on a consultative workplace strategy to gather feedback from all staff who would be using the space. The feedback indicated an overarching desire to work collaboratively but also to enable social gatherings. While this set the planning principles for the fit-out, what was still missing was a conceptual vision for the project.

It was at this point the design team considered the simple and inspirational cookbook by Ferran Adria, The Family Meal. Inspired by the dishes eaten everyday by the staff at the

项目名称：Woods Bagot Melbourne Studio / 地点：Mezzanine, 498 Little Collins Street, Melbourne / 建筑师：Woods Bagot
项目团队：Bruno Mendes - Design leader, Tom Nelson, Brett Simmonds, Richard Galloway, Christopher Free, Sarah Ball
有效楼层面积：200m² / 竣工时间：2016.1 / 摄影师：courtesy of the architect (©Peter Bennetts)

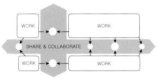
新合作基础设施
a new collaborative infrastructure

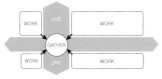
新社交中心
a new social heart

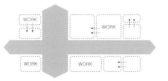
灵活的团队组团
agile team clusters

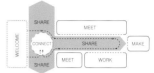
新的客户体验
a new client experience

legendary El Bulli restaurant, the book focuses on the most important time of the day – when the chefs come together to socialise, share ideas and collaborate.

Woods Bagot Principal and design leader Bruno Mendes said the project deployed a sense of domesticity and familiarity. "The simple notion of coming together to share an experience drove our design concept to develop a central heart for the studio. We wanted to celebrate and foster the studio's collective, collegiate spirit."

Based on the notion of gathering, the various formal and informal spaces within the studio were formed around a large central space at the core of the floor plate. Stitching together a variety of informal and formal workspaces, it also forms the main assembly space functioning as an open auditorium by connecting the two levels of the studio while providing a platform for invited speakers, movie nights and dancing on Friday nights. Its connection to the main staff lunch area and through to workspaces was critical to reinforce the social and collaborative agenda.

Monumental in gesture and application, timber was the material of choice. Defining each of the key spaces within the studio, the timber weaves itself into natural steel plates which are used as transitional elements to define the various level changes. The physical result is a studio which is more akin to the nuances of "living" as opposed to "working".

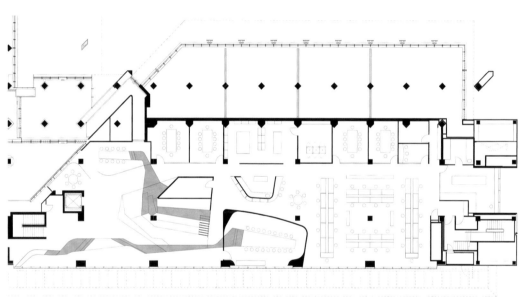

夹层 mezzanine

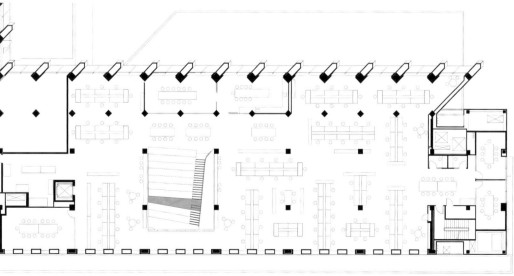

一层 Level 1 floor

建筑师通过增加直角模式的不规则区域来布置楼层平面，使空间显得更加简单明了。

在这个项目中，工作空间被设置在自然采光充足的区域。为了使平面比例协调，立面都很简洁，其中一个立面面对狭窄的街道，从而形成了一个缺乏采光的室内空间。

两层的层高使建筑师有机会更加关注楼层平面的设计。在拆除二层的部分结构后，建筑获得了更好的采光与通风以及一个更公共的室内空间。这种改动相当于为了治病而做的切除手术。

建筑师运用直角模式，给常用的空间更多自然光，并清空二层结构，通过原有的墙体和柜台，安排了大厅和顾客服务空间，将公共空间与私密空间（公司内部使用）区分开，建筑师提倡公共空间与私密空间的共存。石墙的修复与展示只用了一种材料——木材。使用木材是因为它有机、传统的特性，它与天然石材相得益彰。增添木材是将它与石墙融为一体的一种方式：增加不同的元素却不让它与石墙接触。与石墙不同，木板条不是承重墙，而是一种遮盖物，透过它人们可以看到石墙。水平放置的木板条将人引向等待区，而垂直放置的木板条将人的视线引向连接两个楼层的双层空间。

New Arquia Banca Office in Girona

We arrange a floor plan that comes from adding irregular areas in an orthogonal pattern that will organise the new space in a clear and simple way.

This project is designed for workplaces in areas that have more natural light. Floor plan proportions with a minimalistic facade with one of the facades facing a narrow street produce an interior space with scarce light.

The two-story height gives us an opportunity to pay more attention to the floor plan. Gutting and tearing down part of the first floor structure allow for more light and ventilation, in addition to more public interior space. These interventions are similar to surgical cuts, to cure.

赫罗纳Arquia Banca新办公室

Javier de las Heras Solé

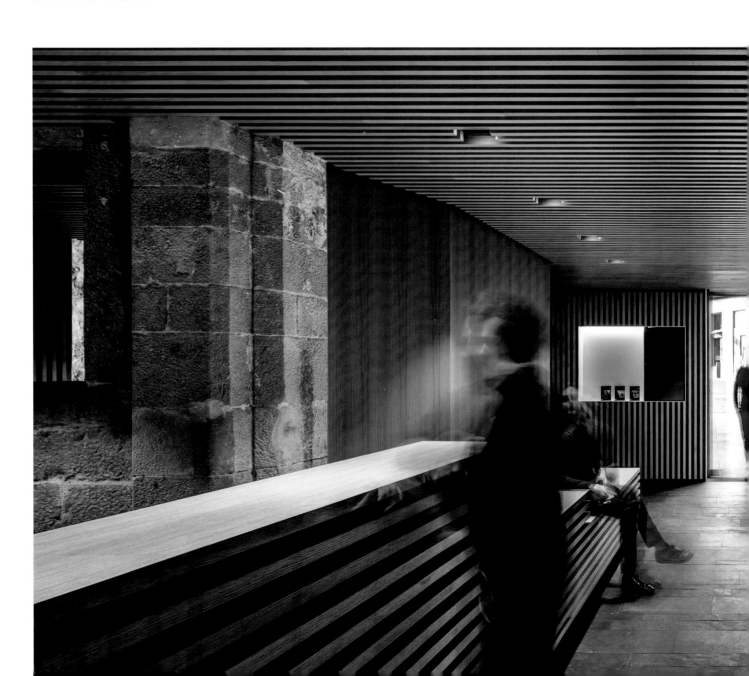

项目名称：New Arquia Banca Office in Girona / 地点：Plaça del vi nº 4, Girona / 建筑师：Javier de las Heras Solé / 项目团队：Salvador Bou Gracia / 结构：Blàzquez-Guanter arquitectes consultors d'estructures / 工程管理：Josep Masachs _ Proisotec Enginyeria / 预算/安全：Gerard Codina Mas / 现场监理：Javier de las Heras Solé / 项目监理：Gerard Codina Mas / 开发商：Montse Nogués Teixidor, Maite Gimeno Pahissa, Berta Toyos Jiménez _ Arquia Banca / 承包商：Joan Illa Casellas _ Construcciones Viladasens S.L. / 木工：Fusteria Guixeres, Joan Guixeras / 锻工：Ramón Cabratosa / 建筑面积：185.8 m² / 设计时间：2014.1 / 施工时间：2014.7—2015.2 / 摄影师：©Adrià Goula (courtesy of the architect)

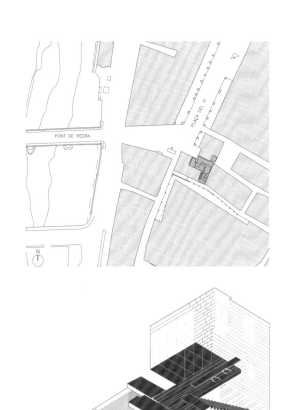

By applying the orthogonal pattern, making natural light a priority for the most-used spaces and emptying the first floor, we can separate public from private space (for the company's internal use) using existing walls and the counter desk disposition that organises the hall and customer care spaces. We propose a coexistence between public and private space. Recover and show the stone walls and use only one material: wood. Wood is a material that because of its organic and traditional characteristics is a good complement for natural stone. Adding this material is a way to integrate it with the existing walls by adding the elements without having them contact. Unlike stone walls, the sequence of wood strips it is not a load-bearing wall, but a covering and that is the way it is revealed. A horizontal wood strips disposition shows us the way in the waiting area, and the vertical disposition leads our gaze towards the double space that connects the two floors.

Javier de las Heras Solé

项目名称：St. Jerome 17 Office
地点：Calle San Jerónimo, 17. Granada, España
建筑师：Javier Castellano Pulido, Tomás García Píriz _CUAC Arquitectura
合作者：Álvaro Castellano Pulido-architect; Fernando Álvarez de Cienfuegos-graphic designer;
Elena Doss, Marta Delle'Ovo, Maria Encarnación Sánchez Mingorance, Alessandro Remelli-architectural students
施工方：Jorge Calvo, Leonardo Cena, GRUPO INNOVAHOGAR DEL SUR, S.L.,Miguel Segura S.L.
技术建筑师：Miguel Ángel Jiménez Dengra
楼面面积：146m² / 工程造价：21,000 Euros (143 Euros /m²)
竣工时间：2015.12
摄影师：©Fernando Alda

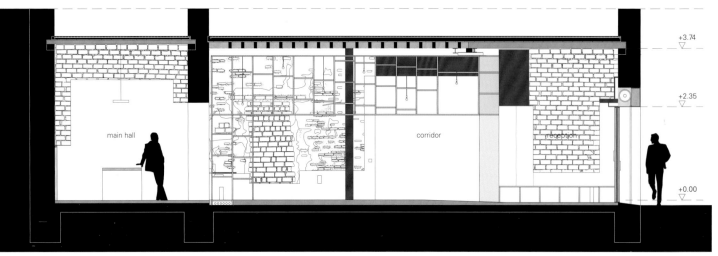

A-A' 剖面图　section A-A'

B-B' 剖面图　section B-B'

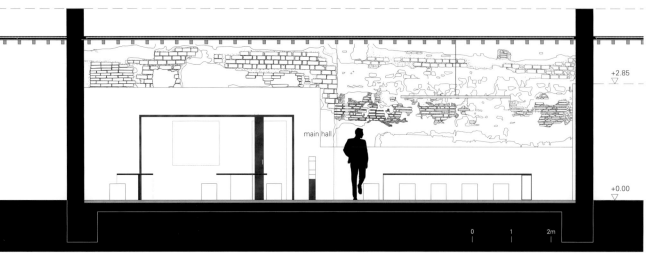

C-C' 剖面图　section C-C'

60 cm thick and wooden floors from the late 19th century, this place is a palimpsest of successive interventions that we transform with recycled elements: a series of shuttering wood pieces taken from a music school work is used for the creation of a channeling-cabinet infrastructure for network cabling and storage of books or models; six wooden doors, some metal shutters and pieces of glass saved from its demolition with several metal profiles from the refurbishment of a house in Granada are assembled for the formation of new holes. Even the plasterboard fragments left without starting by the previous tenant are connected and transformed into a new infrastructure for electricity and lighting.

A 4x1 meters high door taken from our old studio is finally transferred as a cornerstone. San Jeronimo 17 is a project born of opportunities, made of what we find in the place, with the movement of materials from previous works or even with the discovery of unexpected historic contiguities. It is possible to make visible this dynamic, as well as reveal their different strata, mapping and modeling each brick, her wounds, dignifying its heritage presence as part of a continuous history of overlapping elements that we incorporate both minimizing energy invested as our presence.

绿色和平北京办公室
Livil Architects

项目名称：Greenpeace Beijing Office
地点：Dongsishitiao, Beijing, China
建筑师：Liu Chongxiao_Livil Architects
设计团队：Liu Chongxiao, Wang Shilang, Liu Shengpeng, Li Wen, Xu Jingzhang, Li Jianxin
结构设计：Liu Shengpeng
电气专家：Yu Feng, Liu Cheng
设备专家：Meng Xiangkun, Li Yang
客户：Greenpeace
有效楼层面积：1,700m²
建造时间：2015.2—2015.5 / 竣工时间：2015
摄影师：©Wang Hongyue + Lu Wenhui (courtesy of the architect)

绿色和平新办公室位于北京东二环路西侧一个安静的小院内。办公楼建于20世纪60年代，是一个砖结构配木框架的四层建筑，在20世纪80年代及21世纪前十年间经过两次改造，增加了混凝土壁柱和新的钢结构，创建了新的室内空间和室外露台。绿色和平北京办公室占用了三、四层东侧部分和顶部夹层。新设计的目标是使许许多多大小难以精确定义的新旧片段融为一体，形成若干"恍兮惚兮"的不同场景。

部分被粉刷成白色的砖墙和混凝土壁柱经过角磨机和抛光刷抛光，露出本来的面目。当然，打磨工艺也成为一种新的痕迹被带入新材料中。通过除去一些原有的装饰表面层和材料，界定出建筑发展历史中的空间，使建筑变得更容易被人理解。

因切割了焊接钢板而变得轻盈的H型钢柱以及被涂成黑白色的压型钢板天花板使原有的结构流畅过渡到安装了家具的新空间。制作家具的材料包括回收的物流木质托盘，各种木质家具、室外木地板以及木模板、枕木等建筑剩余回收木料。拆除原有的木块和金属板后，这些基本元素经过重新组合以家具的形式重新被安装在墙体中，展现出了完全不同的景观。回收的木托盘也通过重新切割形成了多种不同尺寸的结构体，与金属角钢组合成螺栓状灯具、桌子、座椅，形成了新的空间分隔。

光线在空间布置中起到了唤醒材料本身特征的作用，形状与材料在空间中混合，共同形成一大亮点。空间的流动性通过木材的界面被重新定义出来，不同材料在不同工艺下失去原有的身份，转而形成片段化的屋顶、墙体、门窗以及桌面和柜子。这些片段相互组合穿插，但又不是以传统的物体定义方式，因而显示出某种模糊关系下的多样性。在组合中，这些碎片定义了一个相互分散但又相关的过程——象征绿色和平的发展进程。

Greenpeace Beijing Office

The new Greenpeace office is located in the west of Beijing East Second Ring Road, a quiet and small courtyard. The new brick structure (a wooden frame built in the 1960s) is a 4-storey building, which was reinforced twice in the 1980s and 2000s. After increasing the concrete pilasters and the new steel structure, the new interior space and outdoor terrace were created. Greenpeace Beijing office occupies the east side portions of the third and fourth floors and the top layer ("sandwich"). The goal for the new design is to make the new and old fragments vary in many difficult-to-accurately-define spaces through the integration of each other's way, forming a number of "indistinct indistinct Xi Xi" different scenes.

Parts of the original whitewashed brick wall and concrete pilasters putty covered by iron angle grinders and polishing brush reveal the the original features. Of course, the grinding process has become a new mark into the new material exhibits. By removing some of the existing decorative surface layer, the original presentation of the material, and thus marking space in the history of the building's evolution, the real estate becomes readable.

The H-shaped steel column decreased by cutting off the welding steel sheet, and the pressure plate ceiling painted white and black, make a distinctive transition from the exist-

基本结构
basic structure

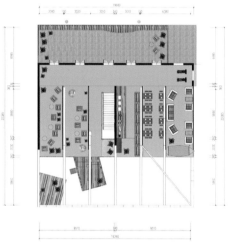

夹层 mezzanine floor

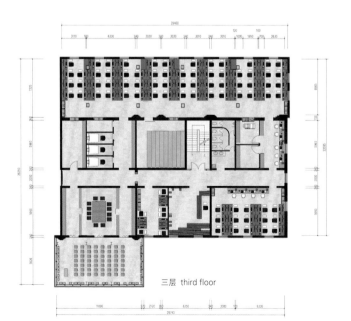

三层 third floor

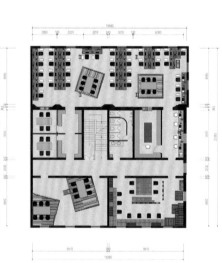

四层 fourth floor

ing structure of the property into a space for furniture: using logistically added wood materials from recycled wooden pallets, many kinds of wooden furniture and outdoor wood flooring, recycled timber from construction remains such as wooden moldings and sleepers. After dismantling the original formation of a block of wood and sheet metal, these basic elements composition re-entered the walls as furniture, which presents a completely different perspective. A stream of recycled wooden pallets were re-cut to form a variety of structures of different sizes: they are combined to form screw-like lamps supported by metal angle irons, tables, and seats, thereby creating the new spatial separation.

Light enters the arrangement to awaken the characteristics of the material itself, while the shapes and materials are mixed

together in space that fosters and becomes Enlightenment. Mobility space has been redefined by the timber interface and various stressed materials lose their identity in a different process, driven in turn to form segments of the roof, a wall, a door and window frames, as well as desktops and cabinets. These fragmented elements have contact with each other – though not in a traditional object-defined style – suggesting a new relationship of somewhat obscure diversity. In combination, these fragments define an interspersed, yet relational, process – a progression that symbolizes Greenpeace.

ACDF建筑事务所为位于蒙特利尔的蓬勃发展的销售终端软件开发商光速公司设计了全新的全球总部，使得几乎被遗忘了的Viger火车站及其旅店重新焕发了生机。这是一座城堡式历史建筑，共三层，其尖顶塔楼俯视着著名的市民广场。ACDF建筑事务所保留了取之于自然的曾经废弃的原始元素，并在原来粗糙原料上叠加了一层极为光滑、充满设计感的元素，反映出光速动态、创新、富有活力的品牌形象。

光速公司自2005年创立以来增长迅速、大胆，从以家庭为基础的商业发展为蒸蒸日上的在欧洲和北美拥有众多分公司的全球实体。公司既想拥有能增加员工人数的空间，同时还要保留初创公司紧密团结、敏捷灵活的企业文化。

自始至终，ACDF建筑事务所都将过去和未来创造性地联系在一起，体现了办公场所独特的传统和生活乐趣（将其与硅谷的标准高科技办公室区分开来）。当代干预措施，如光滑的玻璃墙体、色彩斑斓的楼阁、充满活力的家具和图形艺术——圣保罗Arlin Cristiano和蒙特利尔Jason Botkin的壁画，与工业和自然元素的混搭诉说着建筑的悠久历史。

1898年Viger火车站和旅店首次运营时，富丽堂皇、精致细腻的墙壁使其成为当地标志性建筑。经济大萧条的爆发和蒙特利尔向西方的转向导致了它20世纪30年代的衰败。建筑经历了长久失修，包括15年的彻底废弃，ACDF建筑事务所开始对此办公室进行改造，创造了一个与过去完全不同的全新的火车站。ACDF建筑事务所使高耸的双倍高度空间显露出来，其中的巨大木梁在20世纪50年代改建时被遮挡了；保留了21世纪早期去除石棉后的粗糙的砖墙结构；留下了巨大钢梁的末端。

现在，访客一走进大厅，就能明显感觉到环境的变化。从电梯上看，层压板做成的接待处看起来就像是光速公司红白两色的平面徽标——流畅的L形和S形图案盘绕成鲜艳的火焰。从不同的视角可以看出设计变形的特点。几个边缘锋利的体量大胆自信地组合在一起，在周围纹理粗糙的环境中脱颖而出。

类似的幽默感弥漫在其他的空间。例如，ACDF建筑事务所在大厅里安装了三个小木屋形状的会议厅，每一个看起来都像一座迷你的高光泽度的房子，像家也像光速公司以前设在住宅区内的小得多的办公室一样舒适。另一个设计增强了娱乐感：建筑师在会议厅外与其相连的地板和墙体上涂上了永久的"阴影"。更衣室邻近"游泳池"——厨房一侧的公共区域，参考了以前办公室后院的游泳池。浅绿色的区域有青色的环氧树脂地板和一个玻璃钢凳，凳子上面带有充满水的图案。这些都是找家具设计师Etienne Hotte定制的。两件作品使人恍惚感觉置身于最喜欢的旅游胜地的梦幻般的酒吧。

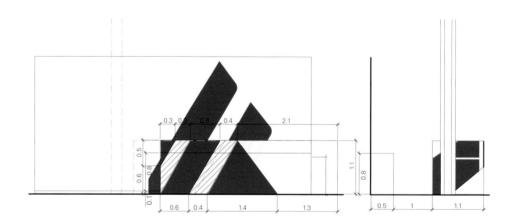

光速公司总部

ACDF Architecture

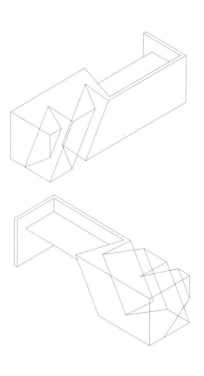

在开放式工作区,保存完好的工业时代的文物,不断升高的天花板,阁楼一样的角落以及优雅散口的砖墙成为朴素的白色系列家具的背景。新旧的对比,Jason Botkin 的图形、抽象线条和形状的注入,这些都能显著促进协同效果,激发年轻员工的想象力,帮助他们实现创新。

Lightspeed Headquarters

For the new global headquarters of Lightspeed, a burgeoning, Montreal-based developer of point-of-sale software, ACDF Architecture reinvigorated three floors of the historical Viger train station and hotel, a nearly forgotten, chateaustyle building, whose pointed turrets overlook a prominent civic square. The studio did so by preserving the found, raw elements of the once-abandoned space, superimposing a layer of select, slick, wit-filled elements that pop against the roughness and reflect the clients' dynamic, creative and vigorous brand.

Lightspeed has had a fast, audacious rise since it was founded in 2005, growing from a home-based business to a thriving global entity with satellites in Europe and North America. At its headquarters, the company wanted room for an expanding workforce while retaining its culture as a tight-knit, nimble startup.

Throughout, ACDF forged connections between the past and the future, reflecting the unique heritage and joie de vivre of its locale (setting it a part from the standard tech offices in Silicon Valley). Contemporary interventions such as slick glass walls, colorful pavilions, vibrant furniture and graphic art – with murals by Sao Paulo's Arlin Cristiano and Montreal's Jason Botkin – juxtapose industrial, found elements that speak to the building's long history.

When the Viger Train Station and Hotel first opened in 1898, it was a local landmark with grandly detailed terra cotta walls.

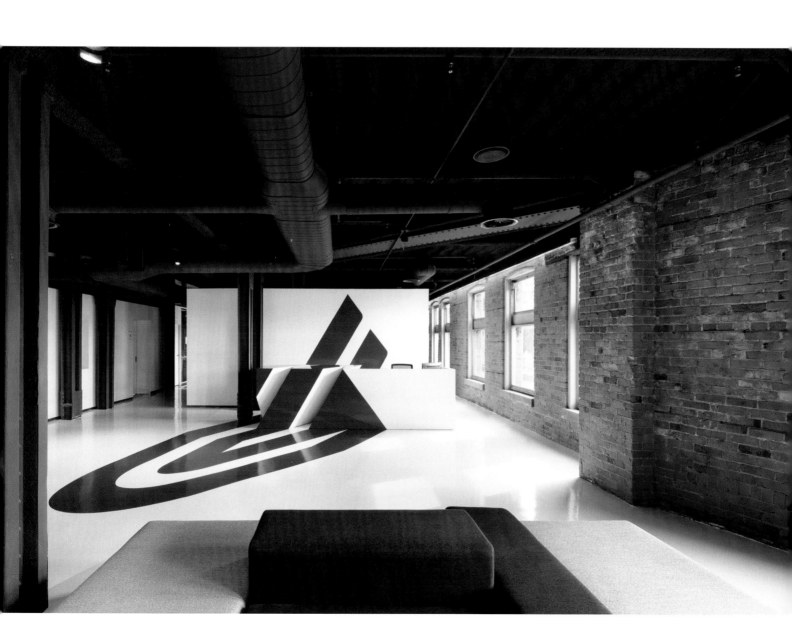

The onset of the Great Depression and the shift of Montreal's downtown to the west caused a downturn in the 1930s. After a long period of disrepair, including 15 years of utter abandonment, ACDF began the office conversion, leaving a palimpsest of the station's past. The studio revealed soaring double height spaces, with their immense timber beams, that were obscured during a 1950s remodeling; retained the rough-hewn brick walls that remained after the structure was stripped of its asbestos in the early 2000s; and left unadorned the ends of monumental, studded steel girders.

Now, a revitalized sense of levity is evident as soon as visitors step into the lobby. When seen from the elevator, the laminate reception desk looks like the flattened graphic of Lightspeed's red-and-white logo – a fluid L and S that coil into the shape of a bright flame. From different perspectives, the anamorphic nature of the design is revealed. An assertive assemblage of sharply edged volumes jumps out, all the more because of the surrounding, coarse textures.

A similar sense of humour pervades the rest of the space. For example, ACDF installed three laminate cabana-shaped meeting pavilions in the lobby. Each looks like a mini, high-gloss house, a nod to the comforts of home as well as Lightspeed's previous, much-smaller office, which was in a residential neighbourhood. An extra layer of wit heightens the sense of play: the studio painted permanent "shadows" on the adjacent floor and walls outside the pavilions. The cabanas are

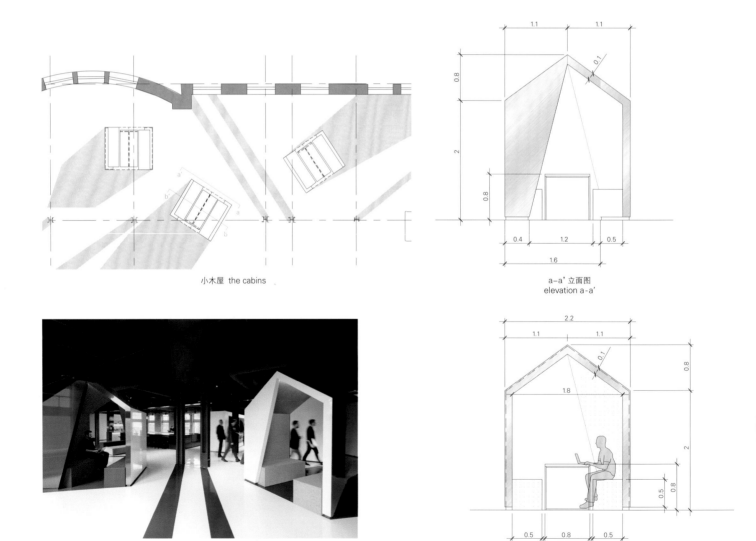

小木屋 the cabins

a-a' 立面图
elevation a-a'

b-b' 剖面图
section b-b'

项目名称：Lightspeed Headquarters / 地点：Montreal, Qc, Canada / 建筑师：ACDF Architecture
项目团队：Maxime-Alexis Frappier, Joan Renaud, Laure Giordani, Laurence Le Beux, Christelle Montreuil Jean-Pois / 机械工程师：Groupe Ce+Co
电气工程师：Scomatech / 项目管理：CBRE / 总承包商：Anjinnov Inc. / 信息技术：Calibre Plus / 视听设备：Environnement Electronique
艺术指导：Speakeasy, Arlin Cristiano, The Doodys, Jason Botkin / 特别家具设计师：Etienne Hotte (pool stools) and Léandre Baillargeon (reception desk)
客户：Lightspeed POS / 有效楼层面积：2,620m² / 设计时间：2013 / 竣工时间：2015 / 摄影师：©Adrien Williams (courtesy of the architect)

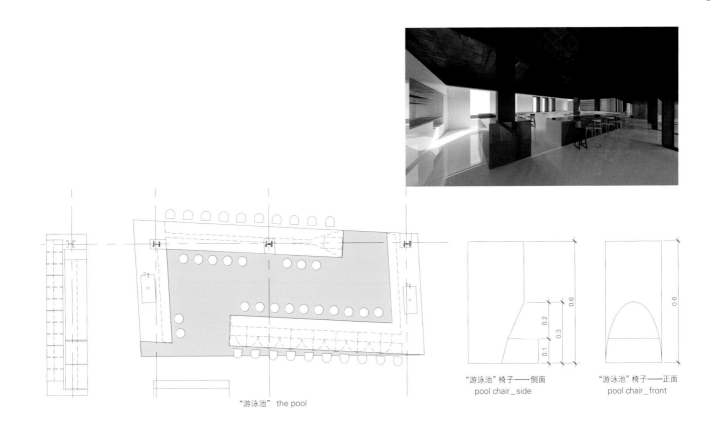

"游泳池" the pool

"游泳池"椅子——侧面
pool chair_side

"游泳池"椅子——正面
pool chair_front

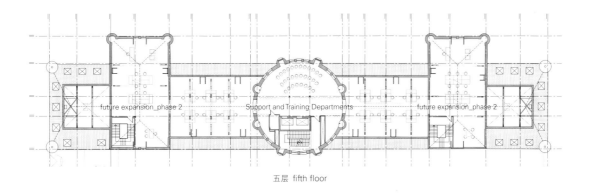

五层 fifth floor

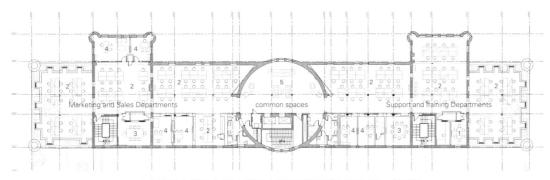

1. 小会议室"休息区" 2. 开放办公室 3. 会议室 4. 封闭办公室 5. 抽烟休息区——第二阶段
1. small meeting room "break-out" 2. open space offices 3. conference room 4. closed office 5. the cigar lounge_phase 2

四层 fourth floor

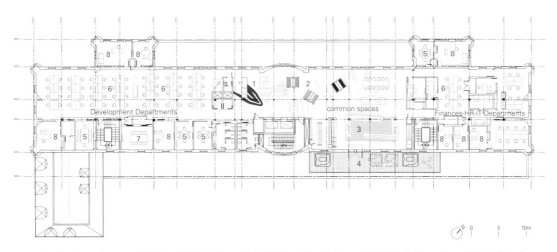

1. 接待处 2. 更衣室——非正式工作区 3. "游泳池"——餐厅与厨房 4. 室外露台——第二阶段 5. 小会议室"休息区" 6. 开放办公室 7. 会议室 8. 封闭办公室
1. reception desk 2. the cabanas_casual work sessions 3. the pool_dining area and kitchen 4. exterior terrace_phase 2
5. small meeting room "break-out" 6. open space offices 7. conference room 8. closed office

三层 third floor

adjacent to the "pool", a kitchen-side common area that references the backyard swimming pool at the previous office. The aqua-colored area has a teal epoxy floor and fiberglass stools with a waterlogged pattern, custommade by millworker Etienne Hotte, both of which make the area feel as dream-like as the bar at your favorite resort.

In the open-plan workspaces, stark white systems furniture streaks past the preserved ruins of industrial-age relics, soaring timber ceilings, garret-like nooks and elegantly frayed brick walls. The new-old contrast, as well as the infusion of Jason Botkin's graphic, abstract lines and shapes, inspires a dramatic synergy that sparks the imagination and helps the youthful employees create.

北京朝阳区的商业园区是创意产业之都。随着大量创业公司入驻园区，同登文化传媒有限公司的海狸工作室也于近期落户其中。工作室所在之处是由之前的一座混凝土厂房改建而成的，MAT设计室将其打造成了一个专为影视传媒公司量身定制的创新工作空间，设计理念是通过情景化的空间体验来诠释其特色和企业文化。建筑师通过增加隔层创造出一个流动空间，从而营造出灵活、互动、趣味盎然的氛围。

原来只有一个体量的桁架结构单层仓库现在需要增加更多的空间感来满足动画、美术、编剧、导演和制片、后勤等基本功能要求。建筑师利用斜顶天花板的高度，通过插入一个多洞口的盒式体量增加了使用面积。这不仅实现了多层空间的互相渗透和流动，同时也为员工创造了一种灵活和互动的气氛。

由于采用的是双层结构，层与层之间的垂直交通流线成为整个空间设计的首要问题。办公室以一种全新的顺序排列，各处皆可沿着螺旋通道实现互动，这也赋予了整个空间以透明感。将五个不同的层次和多孔墙设计融为一体，大大丰富了步行过程中的空间体验。如此，所有区域便真正实现了空间和视觉上的完美结合。

整个空间采用了错层式设计，中庭及所有内部房间，如各自独立的办公室、会议室、排练室和后勤区域，均被安排于这个盒式空间内，而盒子外面的流动空间则留作开放的工作区。公共区域在这个创新的办公项目中是至关重要的组成部分，创意产业的从业者期望在这样的办公项目中能够拥有生活般的、轻松舒适的空间感受。公共区域的一大

北京海狸工坊办公室
MAT Office

亮点在于二层的中庭，阳光可以透过天窗直接洒落到这里。它将一层和二层垂直连接在一起，灯和盆栽植物悬垂于长桌之上，为演讲、公开会议、集思广益和团队建设聚会提供了一个完美的场所。茶水间不是被封闭在一个传统的办公空间内，而是被放大成一个像小岛一样的酒吧，这里可以提供休闲娱乐和上网服务，甚至可以举行小型的临时会议和集体研讨。工作区里，小的休息场所如茶水角或休闲隔间随处可见，它们自然地楔入墙内，营造出一个十分舒适温馨的工作环境。

　　白色的多洞口墙体配以明亮的蓝色作为主色调，使得整个办公空间轮廓鲜明。蓝色图案增加了维度感，同时也为用户带来了新鲜感。黄色和橙色的椅子不仅作为一种补充色，它们还连同反复出现的各种方形框架为整个工作空间注入了活力。清晰简单的设计语言为员工之间的协作和沟通提供了完美的环境。

Beaver Workshop Office in Beijing

The business park in Chaoyang District of Beijing is the capital of creative industries. Along with many other startups in the park, Beaver Workshop of Tongdeng Culture and Media Ltd. has recently settled in a former concrete factory building.

This concrete factory was transformed by MAT office into an innovative work space tailored for a film and media company. The design idea is briefed as contextualized spatial experience to interpret its characteristics and corporate culture. The flowing space is created by adding an interlayer, creating a flexible, interactive, and fun atmosphere.

A single-story truss-structured warehouse, just one large volume, needed more sense of space for the functional requirements of animation, art, screenwriting, directing, producing, and logistics sections. The architects then expand the usable area by inserting a multi-opening box as a second volume, using a ceiling height of sloped roof. This allows multi-layered penetration and flow, and also creates a flexible and interactive atmosphere for employees.

Due to the doubled layers, the vertical circulation between levels merges out as the primary issue in the space. The office has a new sequence where every scene can interact

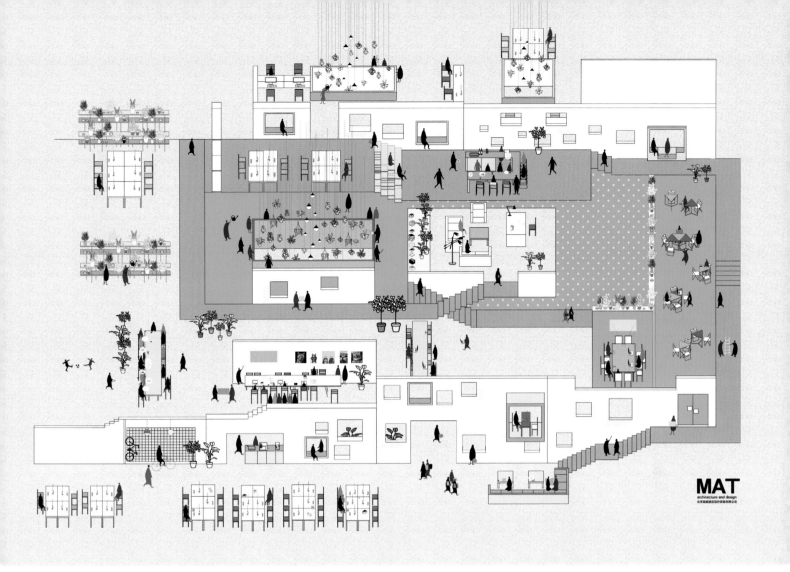

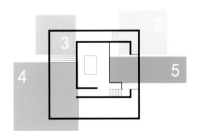

插入式多孔盒子及五个楼层
pluged-in porous box and five floor levels

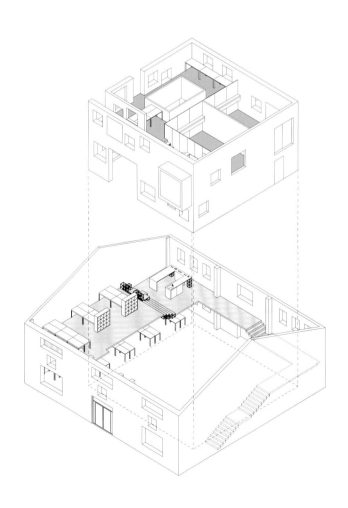

项目名称：Beaver Workshop Office in Beijing / 地点：Sailon Creative Park, Chaoyang, Beijing
建筑师：MAT Office / 设计团队：Kangshuo Tang, Miao Zhang, Siyao Huang, Zhuo Bing, Yanru Li, ZhenXu, Xin Xu
客户：Tong Deng Culture and Media Ltd. / 建筑面积：00m² / 材料：I-beam, concrete, gypsum lath partition, tempered glass, epoxy resin
竣工时间：2016.4 / 摄影师：©Kangshuo Tang (courtesy of the architect)

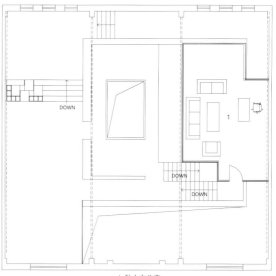

1. 独立办公室
1. independent office
三层 third floor

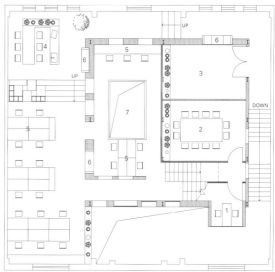

1. 独立办公室 2. 会议室 3. 灵活区域 4. 茶水角/讨论区 5. 工作空间 6. 休息区 7. 中庭
1. independent office 2. meeting room 3. flexible zone
4. tea corner/discussion area 5. workspace 6. resting area 7. atrium
二层 second floor

1. 入口 2. 休息室 3. 会议室 4. 设施 5. 行政管理办公室 6. 灵活区域 7. 茶水角 8. 工作空间 9. 休息区
1. entrance 2. lounge 3. meeting room 4. facility 5. administration office
6. flexible zone 7. tea corner 8. workspace 9. resting area
一层 first floor

with each other along the spiral path, endowing this space a transparent feature. It enriches spatial experiences during walking by the collaboration of five different levels and the porous walls. With this, all the spaces are connected spatially and visually.

A split-level strategy is applied to arrange the spaces. An atrium and all the inward rooms as independent offices, meeting rooms, rehearsal room and logistical spaces are contained in the box, while the flowing space outside the box are left for open work space. Public areas are vital compositions in this innovative office project, and practitioners in creative industries would expect comfortable and relaxing spaces with life-like experience in this kind of office projects. The highlight of public space is the atrium on the second floor with sunlight flooding in through the skylights. It connects first and second floor vertically, with lamps and potted plants hanging above a long table in order to serve a perfect place for presentation, open conference, brainstorming, and team-building party. The tea corner, instead of being enclosed as in a traditional office space, is enlarged into an island bar, providing a rec-

1. 灵活区域
2. 工作空间
3. 中庭
4. 灵活区域

1. flexible zone
2. workspace
3. atrium
4. flexible zone

A-A' 剖面图 section A-A'

1. 休息室
2. 行政管理办公室

1. lounge
2. administration office

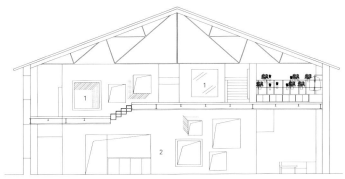

B-B' 剖面图 section B-B'

1. 休息区
2. 工作空间

1. resting area
2. workspace

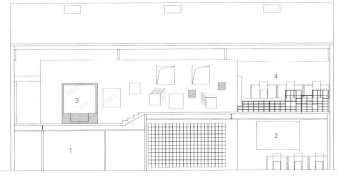

C-C' 剖面图 section C-C'

1. 行政管理办公室
2. 工作空间
3. 休息区
4. 茶水角

1. administration office
2. workspace
3. resting area
4. tea corner

D-D' 剖面图 section D-D'

1. 工作空间

1. workspace

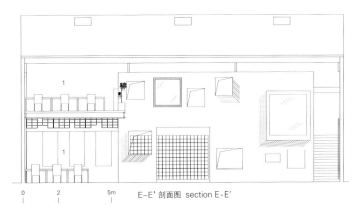

E-E' 剖面图 section E-E'

reation spot and networking platform as well as a space for temporary small meeting and brainstorming. The small places for break such as tea corner and relax cabin can be found among work places, naturally inserted into the wall. This makes a cozy work environment.

The office is defined by white multi-opening walls with as bright blue as an accent color. Blue patterns add a sense of dimension, and also bring freshness to users. Not only yellow and orange chairs as a complementary color but also repetition of various square frames animate the workspace. The clear and simple design language provides the perfect environment for collaboration and communication between employees.

utopic_US共享办公空间，孔德卡萨尔
Izaskun Chinchilla Arquitects

研究表明，共享办公方式已逐渐成为都市生活中不可或缺的一部分。繁华的大都市聚集了无数人，也有了无数的机遇，成为灵感创意的分享之地。因此，以设计向城市致敬并从城市获取灵感是再自然不过的事情了。

utopic_US 是一家年轻而极富创造力甚至可以说有些天马行空的公司，他们的理念是将来自两座城市的灵感与记忆同公司的精髓理念糅合在一起。同时，Izaskun Chinchilla 建筑师事务所通过能够互动且出人意料的设计，来加强空间设计与使用者的联系。

位于孔德卡萨尔的这个办公空间是诞生于此概念下的第一个工作空间，也是事务所选择纽约与东京这两座经典都市的原因。纽约和东京都是让人心生向往、流连忘返且举世闻名的国际大都市，是全世界设计灵感的来源。纽约是开放城市的象征，人们在此开拓创造并取得成功；而东京则代表着最新科技与古老文化的相融交汇。

一代代的年轻人不断推动马德里传统文化的更新。当今，千禧一代已经让我们对网络的重要性与共享的优势深信不疑，而他们现在又一次开始改变马德里。utopic_US 是一个雄心勃勃的项目，作为共享办公空间网络的一部分，它试图成为创业之都，成为改变马德里年轻一代的大本营。Izaskun Chinchilla 建筑师事务所愿意成为这代人的支持者。

这是一个室内设计项目。建筑师设计的多功能办公设备可根据需要调整位置，但他们努力为使用者创造独特的工作体验。他们改变了工业化生产的现代家具（将床改造为书桌，上下铺床位改造为视频会议室），同时引入了色彩斑斓的纺织品、陶瓷制品和彩绘壁纸。

utopic _ US Co-working Space, Conde de Casal

According to many studies, the coworking phenomenon is intrinsically associated to the urban lifestyle. The density and diversity of people and opportunities that a big metropolis produces, encourages the apparition of places where its fundamental purpose is sharing creativity. This is why it seemed natural for us that our spaces payed tribute and were inspired by great cities.

utopic_US is a young and creative company, we could even say it is a little bit carefree and the idea was to mix inspirations and memories of two cities, with the real soul of the firm. At the same time Izaskun Chinchilla Architects looks to strengthen the link between design and user by means of interaction and surprise.

Conde de Casal is the first space we open following this philosophy, this is the reason why we appeal to two classics: New York and Tokyo. Two cities where everyone wants to travel to, where anyone wants to go back to, no matter how

项目名称：utopic_US Co-working Space, Conde de Casal
地点：Conde de Casal, Madrid, Spain
建筑师：Izaskun Chinchilla Architects
项目团队：Adriana Cabello, Alejandro Espallargas, Guillermo Sánchez, Jesús Valer
客户：utopic_US
有效楼层面积：900m² / 竣工时间：2016.05~07
摄影师：©Miguel de Guzman + Rocio Romero_ImagenSubliminal

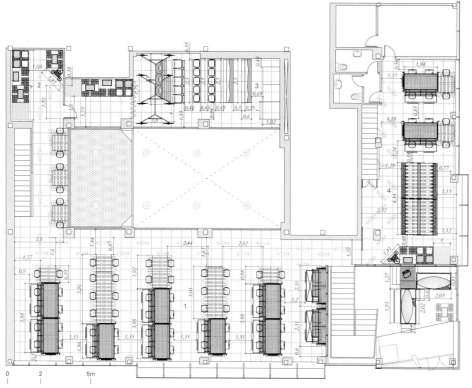

1. 共享办公区 2. 休息区 3. 培训教室 4. 小组讨论室
1. coworking 2. rest area 3. training classroom 4. team room
二层 second floor

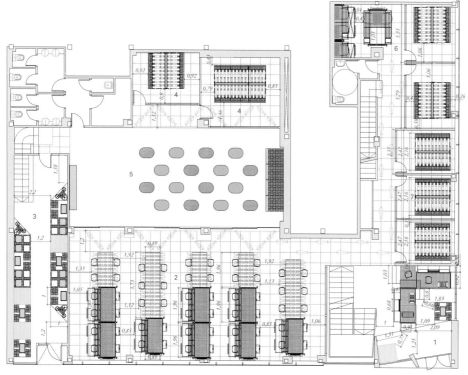

1. 接待区 2. 共享办公区 3. 休息区 4. 会议室 5. 活动区 6. 小组讨论室 7. 办公室
1. reception 2. coworking 3. rest area 4. meeting room 5. event area 6. team room 7. office
一层 first floor

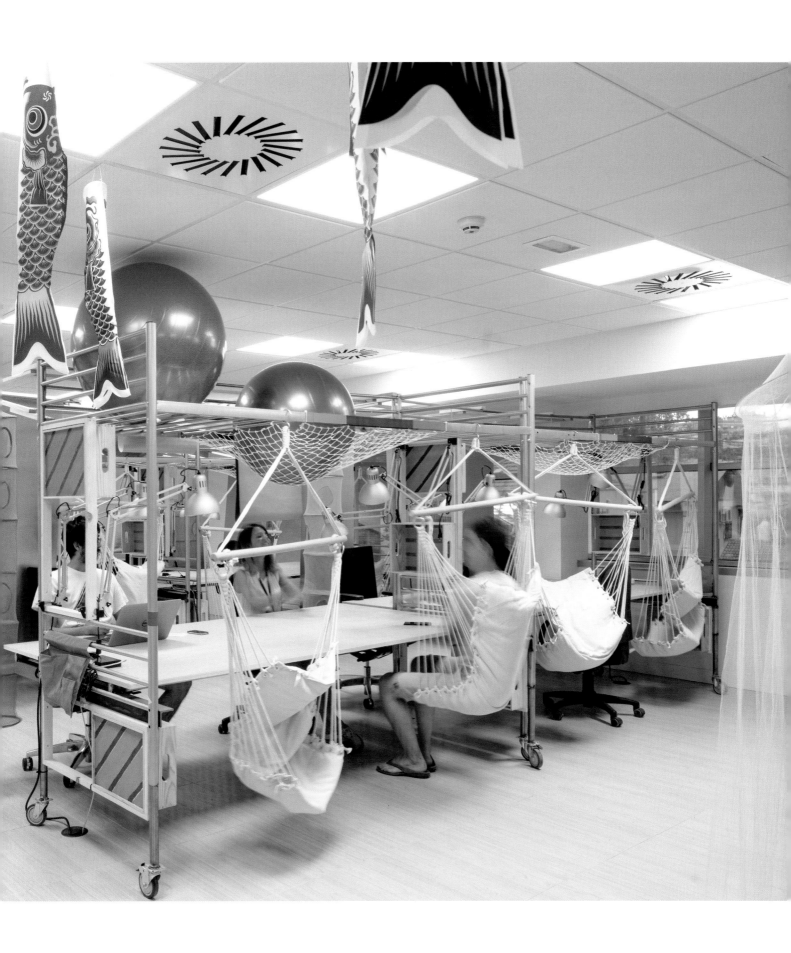

many times you've been to and two cities whose landmarks are recognised all around the globe. They are places that take part of the universal imagination. New York is still the emblem of an opened city where people go to invent and succeed. Whereas Tokyo remains as the symbol of a symbiosis between latest technology and ancestral culture.

Madrid cultural history could be described looking at the transformations that younger generations have imposed. Nowadays the millenials have convinced us of the networking importance and the advantages of sharing. They are starting to transform Madrid once again. utopic_US is an ambitious project, is part of a coworking spaces network that tries to become in the headquarters for this generation to transform Madrid, turning it into the capital of creative entrepreneurship. We want to be the support for that generation.

It is an interior design project. We have designed a versatile equipment that can be moved to another location but trying to create something memorable. We have worked modifying and transforming industrial and contemporary furniture (beds into tables, bunk banks into Skype rooms). We have also introduced a lot of color employing different fabrics, ceramics, painted papers, etc. Izaskun Chinchilla Arquitects

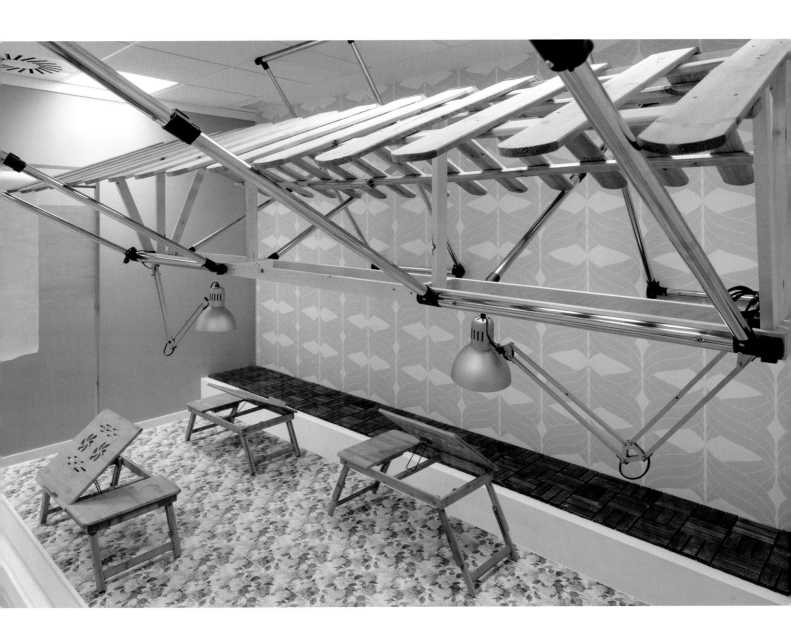

Macro Sea 事务所 2012 年开始对荒废了数年只剩下空壳的 128 号建筑进行改造。设计团队的灵感来自教堂的钢桁架结构，建筑师将其处理成 20 世纪 70 年代高科技现代主义冥想风格，使建筑有一种结构表现主义美感。

这个时期的风格引人入胜之处在于它强调不确定性、环境保护和合作的科技乐观主义，赋予工程、科学和技术以特有的魅力。这个时期同时给我们提供了思索发问的空间：如何才能为今天的设计者、思想家和企业家创造出能激发灵感的当代办公空间？Macro Sea 事务所和 Marvel 建筑事务所一起，创建了各种各样的工作空间、办公室、私人工作室和顶楼寓所以及共享的舒适便利空间，如休息室、公共工作台、先进的原型制作商店和会议场所。建筑团队利用建筑原有的门式起重机设计了吊桥，将社交和展览空间连接起来。他们在保留原有空间表现手法的基础上从原建筑细节中获取灵感对其进行重建和扩建。

新实验室刻意保持了黑白风格，给家具、景观和植物装饰带来的一系列相互依存的高饱和度 Kubrickian 色彩一个呼吸空间。

新实验室
Macro Sea

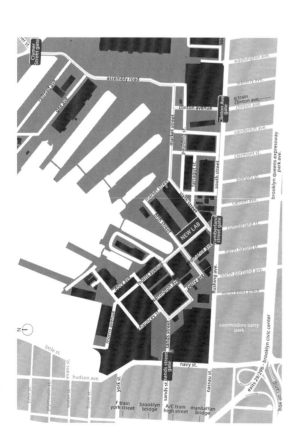

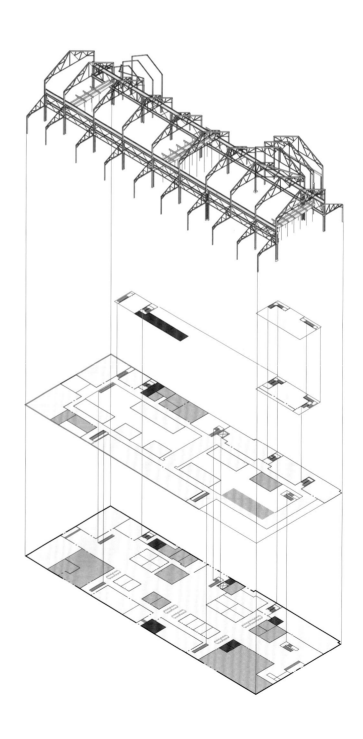

Macro Sea 事务所没有采用开放的楼层平面、传统的格子间和标准的办公家具，而是使用了定制的轻便可拆卸的工作台，用毛毡和木板进行听觉和视觉隔离。个人和小组的格子间与公共工作区和两层楼高的室内广场相互穿插交织，凸显了 Macro Sea 事务所的设计意图：在设计和建造中兼顾保护隐私需要和合作效益。

Macro Sea 事务所采购后现代主义艺术家 Gaetano Pesce、Carlo Mollino 设计的家具的同时，也自主开发了一系列支架式家具，包括 LED 无限距镜照前台，展示实物和数码内容的橱窗以及垂直的景观装置。所有定制家具都在布鲁克林生产组装，大部分就在海军船坞内完成。

New Lab

Macro Sea began work on Building 128 which, after years of neglect, was a deserted shell in 2012. The team took inspiration from the cathedral-like steel trusswork, and approached it as a 1970's High-Tech Modernist musea kind of structural expressionist beauty.

The stylistic period appealed to the design team because of its emphasis on indeterminacy, environmentalism, and a collaborative techno-optimism it conferred a sense of glamour to engineering, science, and technology. The period also lent an appropriate frame through which to ask: How do we create an inspiring, contemporary workspace for today's designers, thinkers, and entrepreneurs? Working with Marvel Architects, Macro Sea created a variety of workspaces, offices, private studios, and lofts with shared amenity spaces like lounges, communal worktables, advanced prototyping shops, and meeting spaces. The architectural team devised suspended

bridges from the building's existing gantry cranes to create connective social and exhibition spaces. The teams took detail cues from the existing building both restoring and expanding upon the inherited language of the space.

New Lab's architecture was deliberately kept black and white to give breathing room to an interdependent range of highly saturated Kubrickian colors deployed through the furnishings, landscape and plant installations.

Eschewing open floor plans and traditional cubiclesas well as typical office furniture Macro Sea custom-designed lightweight, demountable workstations with quilted wool and wood panels for sonic and visual privacy. Secluded spaces for individual or small group work are interspersed with communal work areas and interior plazas over two floors, thus emphasizing Macro Sea's intention to strike a balance between the need for privacy and the benefits of collaboration in design and fabrication.

While sourcing furniture by such late-period Modernists as Gaetano Pesce and Carlo Mollino, Macro Sea developed a range of trestle furniture including an LED infinity-mirror reception desk, exhibition vitrines for physical and digital content, and vertical landscape installations. All custom furniture for the project was fabricated in Brooklyn, largely in the Navy Yard itself.

项目名称：New Lab
地点：Brooklyn, New York, USA
建筑师：Macro Sea
项目管理：DBI Projects
项目建筑师：Marvel Architects
建筑照明：Domingo Gonzalez Associates
古董家具：Ricky Clifton
植物：John Mini Distinctive Landscapes
新实验室合作方：David Belt, Scott Cohen
建筑面积：4,180m²
有效楼层面积：7,432m²
设计时间：2015—2016
竣工时间：2016
摄影师：©Spencer Lowell (courtesy of the architect)

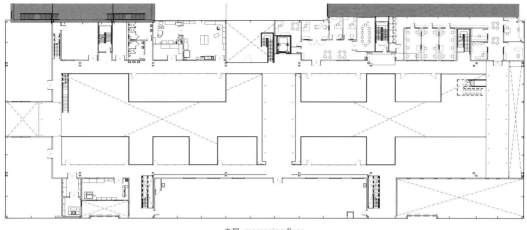

夹层 mezzanine floor

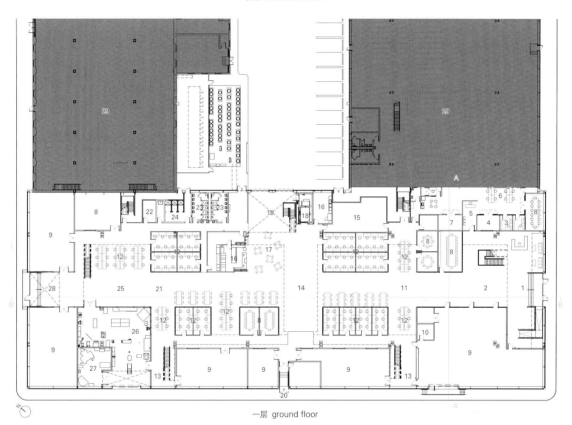

一层 ground floor

1. 入口	13. 楼梯休息区	26. 工作坊	7. lab	19. loading
2. 南展厅	14. 门架展厅	27. 铣削室	8. conference room	20. side entry
3. 经理室	15. 办公室	28. 北入口	9. studio	21. north aisle
4. 展示单元	16. 厨房	29. 相邻租户不在设计范围内	10. green room	22. electrical
5. 前台	17. 咖啡厅座椅		11. south aisle	23. restroom
6. 开放式工作空间与会议空间	18. 电梯	1. entry	12. coworking	24. showers
7. 实验室	19. 装货区	2. south exhibition	13. stair lounge	25. project space
8. 会议室	20. 侧入口	3. manager	14. gantry exhibition	26. workshop
9. 工作室	21. 北通道	4. demonstration unit	15. office	27. mill room
10. 温室	22. 电气室	5. reception	16. kitchen	28. north entry
11. 南通道	23. 卫生间	6. open work and meeting space	17. cafe seating	29. adjacent tenant not in scope
12. 共享办公区	24. 淋浴间		18. elevator	
	25. 放映室			

A-A' 剖面图 section A-A'

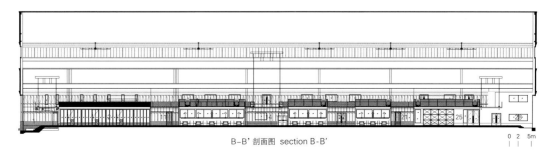

B-B' 剖面图 section B-B'

Rigot Stalars 老厂的修复与扩建
Coldefy & Associés Architectes Urbanistes

该项目位于敦刻尔克市入口铁路沿线的一个老旧的工业园区,离主火车站距离很近,因地理位置好,因此发展迅速。该建筑原为棉纺厂,建于1928年,但已经废弃好多年了。

该建筑曾整修一新,作为现代化合作办公场所,里面有工作室、会议室以及一个小型托儿所和一个自助餐厅。四个功能空间独立运作但共用一个入口。项目的主要目的是保留原有建筑的良好品质,同时改变空间以实现新的功能。同时还要考虑到作为附近街区社交中心的对面广场。

毗邻的水景花园也对项目的设计有类似的影响,因为这是海湾自助餐厅窗口能看到的全景的一部分。为了让此建筑恢复原有的优势,项目完全保留了原有的建筑,并在其地基上增建了两个新的一层空间作为自助餐厅和托儿所。

这次扩建通过可看到周围风景的飘窗和矿物广场使该建筑向远处的城镇敞开大门。矿物广场本身就邻近一处全新的绿色城市空间。这些低矮的建筑物不会遮挡俯视它们的纺织厂,反而有助于突出其原有的结构。

建筑中原先的原材料和现在更清新精致的新材料形成了鲜明的对比。为了获得跟该建筑过去初建时在高原上显示的外观效果,建筑师完全保留了其最初的建筑轮廓,用升降电梯取代了原来的工业电梯,但却保留了楼梯。为了容纳所有的设计项目,新增加了另外两个混凝土板,插入原有的混凝土板之间形成二层和四层。原有整齐匀称的厚砖墙上增加了新的洞口,而外面的金属楼梯打破了立面的单调。

建筑师在改建时通过材料的选择,让玻璃、金属与原来的红砖墙完美地融合在一起,以此获得一定的自由改造权。扩建部分通过颜色淡雅洁净的木材,营造出了宁静祥和的氛围。

家具都是精心设计定做的,办公桌、前台以及酒吧都是由回收的建筑材料制作的。有几件家具还加了麻质软垫,从而进一步强化了一贯的设计理念——更新与继承。

Rigot Stalars, Old Mill Rehabilitation and Extension

The project is located in an old industrial park along the railroad tracks at the entrance of the city of Dunkirk, in fact just a stone's-throw away from the main railway station. The area

is thereby extremely well situated and is developing fast. The building itself had been abandoned and empty for years and was originally a cotton mill, founded back in 1928.

The building has been renovated to house a contemporary co-working environment replete with workrooms, a conference space, a micro-nursery and a cafeteria. The four functions can operate independently but enjoy a common entrance. A principal objective of the project was to preserve the great architectural qualities the building possessed and to transform this space to fulfill new functions whilst respecting the square opposite as a social hub for the neighborhood. The adjoining water-garden also had a similar impact on the design project as well as because it is part of the visible panorama viewed from the bay windows of the cafeteria. To give this symbolic building back its strength, the project fully preserves the existing as it was and implants at its base two new ground floor extensions that house the cafeteria and the nursery.

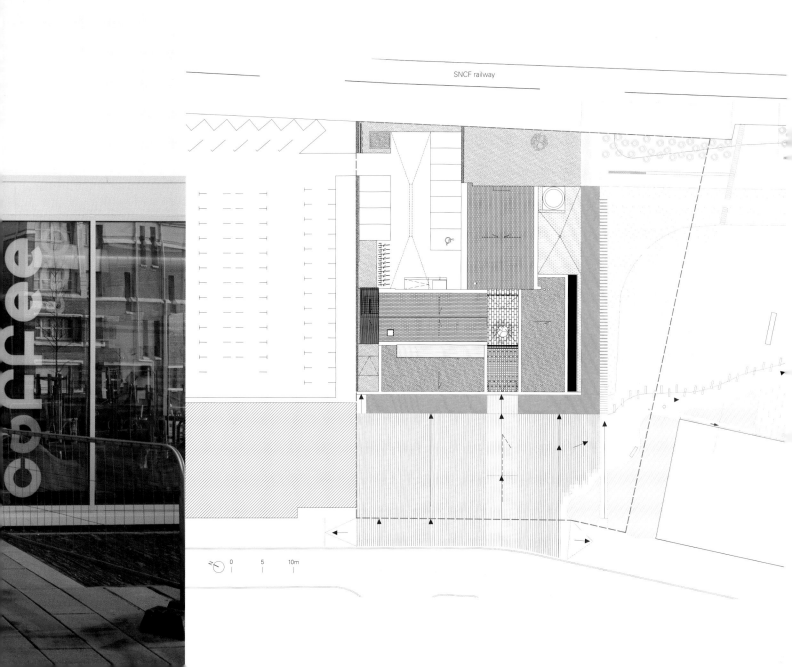

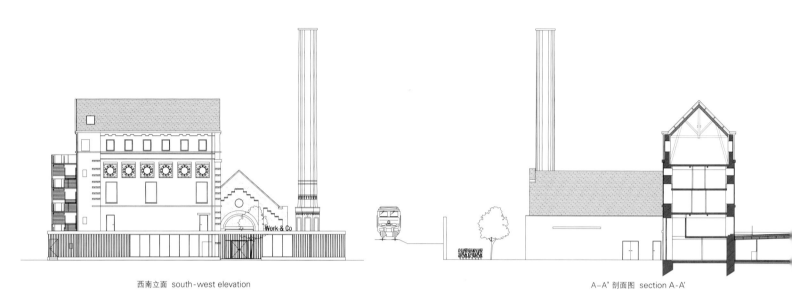

西南立面 south-west elevation　　　　　　　　　　A-A' 剖面图 section A-A'

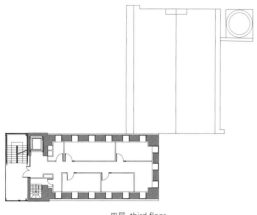
四层 third floor

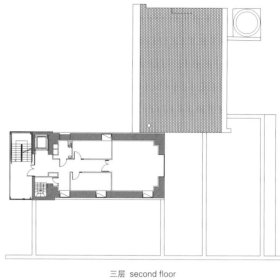
三层 second floor

二层 first floor

一层 ground floor

The extensions open the building up to the town beyond it via the sweeping bay windows and the mineral square, which itself looks on to a new urban green space. These low constructions do not conceal the old mill overlooking them, but actually contribute to highlight the original structure.

In the existing building, the contrast between the raw original materials and the new ones, which are fresher and more refined, creates rather a striking contrast. The original outlines have been entirely preserved in order to obtain the same surface on the plateau as it was in the olden times of its manufacturing days : the elevator has taken over the place once occupied by an industrial lift and the staircase has been conserved. To accommodate the entire program, two additional concrete slabs have been created and just stippled between the existing one, forming level 1 and level 3. New openings in the thick brick walls take place within an existing regularity while the outside metal staircase brings out the singular facades.

A certain discretion is achieved by the choice of materials used in the building's conversion as the glass and metal blend perfectly with the original red brick. A calming, peaceful ambiance has been created thanks to the clear tint of wood with which the extensions were built.

The furniture is designed and custom-tailored: office tables, the reception desk and the cafeteria bar are made from materials recycled from the building's construction and several pieces of furniture have been upholstered with jute fabric thus reinforcing the continuity of the idea of renewal and heritage preservation.

项目名称：Rigot-Stalars, Old Mill Rehabilitation and Extension
地点：15 rue du Jeu de Mail, Dunkirk, France
建筑师：Coldefy & Associés Architectes Urbanistes
环境顾问：Symoe
施工：Norlit
客户：Work & Co, Piet Colruyt (via SCI Social Square)
有效楼层面积：876m²
(rehabilitation-extension of an old mill, the renovated part now houses 479m² of offices and a conference room of 159m², the extension houses a micro-nursery of 117m² and a cafeteria of 121m²)
工程造价：EUR 2,000,000 (before tax)
竣工时间：2014.11
摄影师：©Julien Lanoo (courtesy of the architect)

3XN新工作室

3XN

  3XN原先位于哥本哈根克里斯钦港社区一所历史建筑里，那里的工作空间已不够用了，员工们分散在三个不同的楼层。该公司最近搬到了哥本哈根霍尔曼社区运河边列管建筑"标准船屋"内一个2000m²的新工作室里。此举源于希望有一个空间，大家都可以看到彼此并能随时互动，还可以同时享有模型制作、协同工作的最佳设施以及充足的日光。

  这个历史上著名的标准船屋可以追溯到19世纪中期，最初用于修理和储藏军事船只。该建筑的地面向毗邻的运河微微倾斜，造船工人可以将船滑入水中。该建筑的东立面朝向运河，全部安装了落地窗和活动门。3XN的创始人Kim Herforth Nielsen认为，任何一个人，从高级设计师到实习生，都可以为公司项目和公司的生存发展贡献有价值的想法。他把新工作室设计成一个开放的平面，以工作小组为单位安排员工工作位，便于大家沟通交流。

  竞赛团队坐在一起，项目组、通信组、管理组以及GXN也一样。所有的员工都可以看到彼此，每个小组在做事时可以相互启发灵感。所有员工不分文化等级，都坐在开放的工作室中，培养树立开放和分享的精神。灵活的空间目前安置了80名员工，但这里可以容纳150人。

  3XN在保护建筑外部的同时，拆除了几个原有的内部隔墙以拓展工作空间。他们创建了"盒中盒"，从结构上增加了"自由浮动"的玻璃框会议室，以满足特定的需要。

  Nielsen的目标是保持室内尽可能简单，让公司的工作模型和照片作为公司最典型的特征。白墙、白办公桌以及用玻璃封闭的会议室表

达了这一愿景：处处显露出的具有历史意义的木结构则软化了办公空间的室内设计。

3XN's New Studio

3XN had outgrown its previous workspace, where staff was spread out over three different floors of an historic building in the Christianshavn neighborhood of Copenhagen. The firm recently moved to a new 2,000m² studio in a listed "canon boat house" along a canal in Copenhagen's Holmen neighborhood. The move was inspired by the wish to have one space where all staff could see and interact with each other and benefit from the best facilities for model making, working collaboratively and have ample daylight.

The historic canon boat shed, which dates from the mid-1800s, was originally used to repair and store military boats. The floor of the building slopes slightly towards the adjacent canal, which allowed the boat builders to slide the boats out into the water. The eastern facade faces the canal and is lined entirely with full-height windows and operable doors.

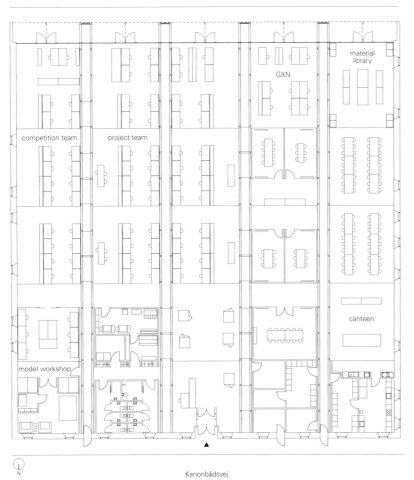

canal

| material library |
| GXN |
| competition team | project team |
| canteen |
| model workshop |

N

Kanonbådsvej

Kim Herforth Nielsen, 3XN's founder, believes that everyone – from senior designers to interns – has valuable ideas to contribute to our projects and the life of the firm. He designed the new studio to facilitate communication, laying it out as one open plan that locates staff based on their group. The competition teams sit together, as do the project teams, Communication, Administration and GXN. All staff can see one another and be inspired by what each group is doing. A non-hierarchical culture, all partners sit in the open studio with the staff, which cultivates openness and sharing. The flexible space currently supports a staff of 80 but can accommodate up to 150.

While the exterior of the building is protected, 3XN did remove several existing interior dividers to open up the working area. They created a "box within a box", adding structurally "free floating" glass-framed conference rooms to satisfy this particular need.

Nielsen's goal was to keep the interior as simple as possible, to allow the models and photographs of the firm's work to be the defining features. White walls, white desks and glass enclosed meeting rooms express this vision. The historic wooden structure is revealed throughout, which softens the interior.

项目名称：3XN's Studio / 地点：Copenhagen, Denmark
建筑师：3XN Architects / 项目团队：Kim Herforth Nielsen, Bo Boje Larsen, Jan Ammundsen, Signe Blomquist, Jeppe Kongstad Hjort
机电管道工程师：Leif Larsen Ventilation / 施工方：HK Byg Enterprise A/S
建筑面积：2,000m² / 有效楼层面积：2,000m² / 竣工时间：2014.12
摄影师：©Adam Mørk (courtesy of the architect)

室内工作环境、办公室和工作室
Working Interiors, Offices and Studios

办公室变革

多年来,工作环境已经发生了改变,譬如,组织原则、工作条件、美学技术、工作空间新文化。并不是每一个人都在办公室里工作,但在不同种类的工作空间里;办公室类型仍然占据主导地位。从前台到办公隔间再到会议室,数个小时的工作经济在办公室里发生。有趣的是,如今的办公室室内设计风格迥异,以至于让人不禁对其概念和定义产生怀疑。关于理想的办公室,人们反复围绕保证最大限度的灵活性、保证合作、保护隐私这三点进行讨论。或许,新型办公室的基本趋势在于对空间的改造性再利用,这符合新业务中产生的文化,而不存在于传统的办公室大楼中。在众多事情发生于虚拟空间的时代,实体办公室可以为在不同文化中工作的人提供日益重要的社会交往的机会。什么是办公室变革?办公室室内环境的进化历史在 21 世纪会如何显现?

办公室与我们现代社会完美地融合在一起,我们却不常想起其发展历史。从全球的角度看来,办公空间这个概念真正起源于西方,是一种复杂的组织机构,在几个世纪以前空间变得更加宽敞。在欧洲,早期的贸易公司开始雇用职员和会计。他们或许是第一批办公人员。同样,许多店主雇用职员,并为他们提供住宿。官僚主义的根基扎得越来越牢固,办公领域亦是如此。早在一个世纪之前,办公室的布局已经发生了改变,从私人的办公室、公共走廊发展到第二次世界大战之前的开放但有秩序的"工作楼层"。如今许多办公大楼的平面布局都可以追溯到路易斯·沙利文于 1891 年设计建造的温莱特大厦。尽管建筑创新性地使用了钢构架,可以实现较大的窗户和较薄的办公室墙体,但温莱

The Office Evolution

The built environment for working has evolved over the years, as organizational principles, working conditions, and aesthetics adopt technology and the new culture in the workplace. Not everyone works in an office but among the various kinds of workspace, the office typology is still the most predominant. From reception desks to cubicles to meeting rooms, many hours of the working economy take place in the office. Interestingly, the designs of office interiors vary so much today that one would question the idea and definition of an office. Discussions about the optimal office have gone back and forth from maximum flexibility to collaboration to privacy, and a few timezs over again. Perhaps, the underlying trend of the new office isn't found in traditional office buildings but rather in an adaptive reuse of space that suits the culture of new businesses. In a time where much takes place in the virtual space, the physical office offers people in a varied working culture an increasingly important part of human social contact. What is the office evolution and how will the history of office interiors manifest itself in the 21st century?

The office is so integrated with our modern society we don't often think of its history. From a global perspective, the office space really stemmed out of a western idea as complex organizations became larger a few centuries ago. In Europe, early trading companies began employing clerks and accountants. They were probably the first office workers. Also, many shop owners had live-in clerks as helpers. As bureaucracy became more established, so did the office. Just a century ago, the development of office layouts has changed from private individual rooms with outside corridors, to an open yet still orderly "work floor" just before the Second World War. Many of the office buildings today still have floor plans that trace back to the Wainwright Building in 1891 by Louis Sullivan. Despite the innovative steel-frame skeleton that allowed larger window openings and thinner interior

温莱特大厦，圣路易斯，美国，1891年，路易斯·沙利文
Wainwright Building, St. Louis, USA, 1891, Louis Sullivan

大工作室，强生公司总部，威斯康星州，1936年，弗兰克·劳埃德·赖特
Great Workroom, Johnson Wax Headquarters, Wisconsin, 1936, Frank Lloyd Wright

新实验室，美国，2016年，Macro Sea
New Lab, USA, 2016, Macro Sea

特大厦的内部布局仍然采用那个时期典型的私人办公室布置。

办公空间这个概念后来一直到1936年弗兰克·劳埃德·赖特在强生公司总部建造"伟大的办公室"时才得到检验。这个办公室是开放的，有双层高的扩展空间，并且内部没有墙体。这本来是为秘书们建造的，让他们与在夹层工作的行政人员一同在开放的平面上工作。

20世纪50年代和60年代，现代主义者开始接受开放性的办公室，把其视为一种工作原则，这种原则因办公室景观而受到欢迎。但是工作人员从未习惯这类办公室，他们渴望获得更多私密空间。1968年，赫曼·米勒公司设计师罗伯特·普罗普斯特发明了"行动办公体系"，它是一种可以重构成开放式办公室的家具系统，这导致了办公隔间行业的出现。

在强生公司总部，从夹层看到的开放工作空间的景象可以使人想起雅克·塔蒂1967年经典影片《玩乐时间》里的场景。然而与赖特的开放办公空间不同，主角Monsieur Hulot从上往下看到布满一个个隔间的迷宫般的办公室楼层，最终也没有找到他要找的会计师。这部幽默影片对作为技术未来一部分的办公室空间与人类之间的关系提出了质疑。今天这一问题在办公室设计讨论中仍然存在。

最近，Marco Sea事务所的开发人员和设计师在将一个造船厂改造成新实验室时对人类与未来这一主题进行了讨论。新实验室是为人工智能、机器人这样的新技术而设计的工作空间。从上往下快速地瞥一眼一个个隔开的工作空间，就像塔蒂场景的一张现代快照。但仔细看就会发现，为了与巨大的教堂风格的空间共存并在其中蓬勃发展，

walls, the Wainwright building still had an interior layout of individual offices typical of its time.

The idea of an office workspace was only tested later when Frank Lloyd Wright created the "Great Workroom" in the Johnson Wax Headquarters Building in 1936. It is an open, double-height expansive space with no interior walls. It was originally designed for secretaries to work in an open floor plan, with administrators situated on the mezzanine.

In the 1950s and 1960s, the modernists embraced the open office plan as a philosophy, popularized by the idea of Bürolandschaft, the office landscape. However, workers never got used to the idea, and longed for more privacy. In 1968, Robert Propst, Herman Miller's designer, invented the "Action Office System", which was a furniture system that was reconfigurable to compliment an open office plan. This set off the industry of cubicles.

The mezzanine view of the open workspace in the Johnson Wax Headquarters is a reminder of a scene in Jacques Tati's 1967 masterpiece film, Play Time. However, unlike Wright's open workspaces, the main character Monsieur Hulot looks from above onto a maze-like office floor full of cubicles, and ultimately failed to navigate through office floor and find a specific accountant. The comical film questioned the relationship of humanity and the office space as a part of a technological future, a question that is still valid in the dialogue of office design today.

More recently, developers and designers of Macro Sea dealt with the very subject of humanity and the future as they repurposed an old shipbuilding facility into the New Lab, a workspace for new technology such as Artificial Intelligence and Robotics. At a quick glance from above onto the divided workspaces, it is a modern snapshot of the Tati scene. But upon a closer look, the various open, leisure, and private workspaces are made to be intimate and vibrant in order to coexist and thrive within the large cathedral-like space.

In the 1980s, while office workers had their cubicles, the

AKQA东京办事处，日本
AKQA Tokyo Office, Japan

光速公司总部，加拿大
Lightspeed Headquarters, Canada

各种各样的开放空间、休闲空间和私人工作空间变得更加亲密和充满活力。20世纪80年代，办公室职员拥有自己的格子办公间时，公司主管却坚持沿着立面和角落设置私密办公室。但在20世纪90年代，管理人员和员工之间的讨论进行到政治层面时，新兴企业却没有营造"企业形象"，而是采用了更加水平的结构，因此，开放高敞的办公空间再一次流行起来。小企业的态度对现在新型办公空间仍然有重要影响。

对于市区或者郊区工业园区的办公大楼，人们试图在老建筑里寻找工作空间。这些老建筑之前有过其他的用途，通常是工业用途，例如，火车站、旅馆、仓库和工厂。今天，在成功的改造项目中，可以看到无数充满活力的高敞空间。debartolo建筑师事务所拆除了于1992年建造的百货商店不必要的室内装饰，只露出必要的基本构件，设计了VSCO奥克兰办公室，塑造了一个年轻、有创造力的摄影公司。将喧闹的功能空间设计在外围、将安静空间设计在办公空间内的规划方法有助于为80人创造一个高效的工作空间。另一个例子就是Torafu建筑师事务所设计的AKQA东京办事处。尽管办公室位于地下室，但是设计师对内部庭院的组织最大限度地利用了自然光线。

建筑结构和通信交流两方面的技术不断进步，灵活的工作场所随之产生。员工可以在任意地方工作，如果愿意的话可以在小隔间，第二天也可以坐在变形椅上工作。一些员工开始是在某些地点办公，后来老板甚至要求他们定期在家里工作，这样公司能最大限度地利用办公场所，与更多的员工共享。工作中这种自由流动打破了以前办公室的规则和功能。

executives insisted private offices along the facade and corners. But when management-employee discussions carried on into a more politically correct 1990s, younger businesses shy away from the "corporate image" and harness a more horizontally ranked structure, and with that, the idea of an open, loft-like office spaces became again widespread. That attitude still carries weight for the new office space today. In reaction to office towers of downtowns or suburban business parks, the new workplaces are seeking a home in older buildings that had a former use as something else, usually industrial: Train stations, hotels, warehouses and factories. Numerous examples of vibrant loft-spaces can be seen in today's successful renovation projects. The VSCO Oakland office by debartolo Architects captures the energy of the youthful creative and photographic company by removing layers of unnecessary interiors of the 1922 department store to reveal only the core essence of what is needed. The approach of planning noisy programs along the perimeter and quieter spaces inside works has helped create an efficient workspace for 80 people. Another office is the AKQA Tokyo Office by Torafu Archtiects. Although the space is in the basement, the designer's organization around an inner courtyard helps maximize natural lighting.

As technology advances in both building construction and in communication, the flexible workspace was introduced. The employee can work anywhere, a cubicle if one desires or on a beanbag chair the next day. Began as a perk to work in certain places, some employees are now even required to periodically work from home so that companies can maximize office capacity to share with more people. This freedom of movement in working breaks the older rules of the office's role and function it provides.

Today, the rule of office interiors is not about having rules. While the fundamental principles of office layouts remain a guide, the interpretation of those principles and aesthetics of the office interiors could be implemented more loosely.

3XN新工作室，丹麦
3XN's New Studio, Denmark

今天，办公室室内的规则并不是制定规则。办公室布局的基本原则已成为一条指导方针，对那些原则和办公室室内结构美观的阐释不必那么严格。尽管在不同的文化和企业中，组织方式各不相同，但都倾向于创造场所，营造合作轻松的氛围，同时平衡好用于生产的场所，使生产效率不受到干扰。由于受到高生产效率和利益的驱动，不同的安排方式、合作空间、照明和其他的工作条件都蕴含了开放性和创新性。同时，因为企业形象和品牌与办公室美观联系在一起，所以企业形象和品牌开始对其产生影响。因为机构组织努力树立声誉，吸引消费者，雇用新员工，所以办公室设计在公共关系中日益成为其"面子"中不可或缺的一部分。软件开发商光速公司几乎从字面上理解这一概念的意思，在新建的公司总部把鲜红色和白色的标识纳入前台设计。一出电梯，公司标识就会玩笑似的"映入眼帘"。红色的线条穿过白色的环氧树脂地面，到达桌子上和后面的墙上，形成一个扁平的、缩小版的标识图形。

ADCF建筑事务所的建筑师为气氛轻松、时尚雅致、专业的光速公司总部设计出独特的具有代表性的工作空间，设有玻璃墙、颜色各异的展台，还有混合的冷色系壁纸和壁画，与改造的废旧火车站和旅馆暖色系砖块形成对比。

在环境和机会合适时，老建筑的再利用可以是很高效的。一个简单的方案就是打通一系列小空间，营造一个大空间。3XN在哥本哈根建立了新的工作室，有机会把19世纪的船屋打造成开放的海滨办公室，长长的毫无阻挡的空间指向水面。即使有的地方需要隔墙，隔墙也采

And while organizations vary in culture and operations, they tend to seek out spaces that foster the ease of collaboration while balancing the spaces for productivity without distraction. The drive for better productivity and profit has created openness and innovation to a variety of arrangements, collaborative spaces, lighting and other working conditions. At the same time, business image and branding comes into play as it correlates with aesthetics of the office. More and more, the office design has become an integral part of the organization's "face" in public relations as it strive to build reputations, attract customers and recruit new employees. Almost taking that idea literally, the new headquarter of software developer Lightspeed has incorporated their bright red and white logo as a part of their reception design. Stepping out of the elevator, the logo is playfully "in your face". The use of red is stretched across the white epoxy floor, and onto the desk and the back wall to create a flattened, foreshortened perspective of the logo graphic.

The architects at ADCF Architecture created a unique representative space for Lightspeed Headquarters that is lighthearted, yet tastefully professional and sleek, with interventions of glass walls, colorful booths, and a blend of cool art and murals against warm bricks of a converted old train station and hotel.

The repurposing of old buildings can be modestly powerful when the context and opportunity is right. One simple scheme is to open up a series of smaller spaces to create a larger one. 3XN's new studio in Copenhagen had that opportunity by turning 19th century boat sheds into an open waterfront office environment with long, unobstructed space pointing towards the water. Partitions, even when needed, are glass walls to strengthen transparency and the open plan.

Another trend that is radically changing the concept of office is the co-working space, especially in urban areas where rent may be unaffordable for many. The opportunity

用了能增强透明度和开放性的玻璃墙。

共享办公空间是另一种彻底改变办公室概念的趋势,尤其是在许多人付不起房租的城市里。和志趣相投的人在一起共享设施这种机会吸引着个体经营者、小型企业和创业者。吸引都市人的一点是他们拥有选择权和自由,可以坐在其他人中间工作,可以在私密场所工作,也可以在会议室里进行电话会议。如同几乎可以在任意一个大城市找到健身房,你可以在任意地点工作,可以通过音频或视频会议与任何人交流。这些新潮办公空间的美感和尝试反映出对较为年轻的、工作时间短暂的员工的吸引力。能体现出这种趋势的项目是由 Izaskun Chinchilla 建筑师事务所设计的 utopic_US 共享办公空间。庸俗廉价物品的巧妙组合为那些愿意在气氛极其活跃的地方工作的人提供了一个有趣的地方。

自然光线和工作环境一直是设计办公室环境的主题。通常来说,想要控制直接照射进工作空间的光线是非常困难的,因为它会产生刺眼的强光或者造成电脑屏幕和会议室光线太亮。同时,自然光线和与外面世界的接触有益于创造健康的工作环境。OMA 事务所对位于纽约的蔡国强工作室的改造和扩建以另一种方式利用了光线。在众多的策略中,设计的目的是让自然光线照进地下室中,创造出愉悦的工作环境。为了达到这一目的,建筑师设计了一系列的采光井。对镜子的灵活运用不仅让间接光照进工作空间,而且可以让员工从地下空间看到大街上的景色。

工作室空间

尽管许多有创造力的专业人员在办公室里工作,但是专业性的办

to share facilities while having like-minded people around is an attractive option for the self-employed, small businesses and start-ups. What is appealing to the urbanite is the choice and freedom to sit amongst others, work in a private space, or teleconferencing in a meeting room. And like a gym, you can find one in almost any large city, allowing one to work from anywhere, and communicate with anyone using audio or video conferencing. The appeal of a younger, transient workforce is reflected in the aesthetics and experimentation of these hip spaces. A project that show this trend is the utopic_US co-working space by Izaskun Chinchilla Architects. The clever assembly of kitsch and cheap objects make this a fun place for those who can work in a highly visually active environment.

Natural daylight and the working environment has always been a design topic in the office environment. Often, it is difficult to control direct lighting into the workspace as it create glare or too much brightness for computer screens or meeting rooms. At the same time, natural lighting and contact to the outside promotes a healthy working environment. OMA's renovation and expansion of Cai Guo-Qiang's Studio in New York explores light in another way. Among many gambits, the design aims to bring natural lighting into the basement to create a pleasant working environment. It does so by the use of a series of light wells. The clever use of mirrors not only brings indirect light into the space but also offers a view into the street from below.

Studio Space

While many creative professionals work in an office, the professional office doesn't always work for the creative. Shying away from a business-like surrounding, some people prefer calling their workspace the studio. More than just a notion of a work place or an office, the studio is a creative space. A space where ideas are born, are developed and tested, and have been given life. It is where fragments of

utopic_US合作办公空间，孔德卡萨尔，西班牙
utopic_US Co-working Space, Conde de Casal, Spain

蔡国强工作室，美国
Cai Guo-Qiang's Studio, USA

公室并不总是为有创造力的人服务的。一些人特意回避商业化的工作环境，更喜欢把自己的工作空间称为工作室。工作室不仅仅是工作空间的概念或者是办公室，更是一个有创造力的地方，各种想法在这个地方产生、成熟、检验、被赋予生命。正是在这个地方，创新思维的零星片段汇聚一同显现出来，并通过各种媒介表达出来，比如：使用油漆还是墨水；通过珠宝还是影片；以模型还是文字的方式呈现。有的思维通过工作室的前门展现给世界，而另一些则被静静地存储起来。工作室空间同样也是深思和生产的场所。在某些特定时刻，存储的文件是深思、重新评估和鉴定的来源。从那个层面上看，studio 和它的拉丁语起源 studium 的意思非常接近。根据研究，其意思是读写和探索的场所。工作室有时会成为展示空间，展现一段特定时间内的发现情况，讲述故事及其制作过程。回顾之前的作品，或者后退一步，去看一下制作中的作品，人们可能受到鼓舞，继续前进。工作室可以使创作者回到当下。创作者生产出用于使用、展示或将来能得到赏识的工艺品，在这一"流动"的过程中，时间再一次流逝。因此，对于在工作室空间办公的创作者或者艺术家来说，工作室空间就是时间胶囊。工作室建筑本身在周围新旧环境下，或许与时间也有联系。但是对于创作者（无论他或她是否是工作室的拥有者）、建筑师、艺术家、摄影师或者电影制作人来说，工作室是庇护所，是安全之地。在这里，人们可以和自己的创作过程与手工艺品一起消失在时间里。夜深的时候，创造力四溢，这时候工作和玩耍之间并没有什么区别。创作者全神贯注，冒着风险，设法完成自己的构想。正如同每一位创作者的创作过程和工艺品都是独一无二的，为了符合和体现使用者的需求和艺术立场，这一部分提到的工作室在重要性、环境、空间组织方面存在差异。

creative thoughts are pulled together, manifested and is together expressed through various mediums: in paint or in ink; in jewelry of in film; in model or in words. Some make it through the front door into the world, and others are quietly archived. The studio space is also a place of reflection and production. On occasional and certain moments, the archive is a source of reflection, re-evaluation, and appreciation. In that way, the studio comes closer to its latin origin, studium, as a study: the space to read, write and to discover. The Studio can at times become an exhibition space that captures that discovery work of a specific time, and tells a story and the process of the making. Upon looking back at older works, or taking a step back to glance at a work in progress, one may be inspired to go further, and the studio space offers the creator to come back and be in the present, and in that moment of "flow", time is lost again as the creator produces one's craft to be used, shown, or again appreciated in the future. Therefore, the studio space is a time capsule to the creators or artists that uses them. The studio building itself may also have a relationship of time with its context of the old and new surroundings. But to the creator, whether he or she is the studio owner, the architect, the artist, the photographer, or the filmmaker, the studio is a sanctuary and safe place where one can be lost in time with his or her own creative processes and craft. In the late nights when creative juices flow, there is no difference between work and play. The creator is focus and takes risks while pushing his or her vision through. Just as every creator's process and craft is unique, the studios in this review also vary in their materiality, context, and spatial organization to match and reflect the user's needs and artistic posture.

Contextually, the Architecture Archive in a Somerset farmyard, England, may not be the first place to look when thinking about studios. However, upon a closer look, this archive, as the name suggests, houses an office and the

建筑档案馆，英国
Architecture Archive, United Kingdom

联系上下文，想起工作室时，第一个要去观看的地方并不是位于英国萨默塞特农场的建筑档案馆。但是，走近一点看就会发现，这个处在农业环境下的档案馆正如其名字所指的那样，里面有一个办公室，还有馆主有关建筑和家庭的收藏品。作为一个保存文件和深思的地方，这个档案馆坐落在农场旧谷仓墙体的后面，十分低调，以此来探索时间元素。得到部分修复的古老砖石墙体诉说着过去的故事，而其背后两个当代木结构暗示了其崭新之处。该建筑规模大小合适，在周围的建筑物中并不突出或者格格不入。总部位于伦敦的Hugh Strange建筑师事务所设计的实木胶合层压墙体结构采用了简单而又大胆的形状，是一个聪明、有效的建筑方法，这不仅有利于稳定档案馆的内部结构，而且通过预制高质量的组件有效地满足了不太充裕的建筑预算要求。拱形的波纹屋顶把结构统一为单一的样式。大量的阳光穿过屋顶的采光口照进档案馆，屋内阳光充足。建筑师在对这个档案馆进行最后修整时，把薄壳结构的简易程度和定制内部结构的个性化进行了调整，使之达到平衡。展示的工艺和对熟悉材料的再利用反映了这个档案馆独特的背景。重新利用附近倒下的树木来建造实木地板，这种态度表现了一种环保意识和对资源的尊重，这样的设计创造了人性化的、舒适的室内环境，成功地建立了一个工作室空间，为思索、发现手边的收藏品提供了理想的环境。缜密的设计使得这个项目获得了多个奖项的提名，包括2015年欧盟当代建筑奖、密斯·凡·德·罗奖、英国皇家建筑师协会国家奖和地方奖。

阿克雅工作室是阿克雅建筑事务所在圣保罗新建的总部。在这个项目中，工作室的概念并不仅仅是工作的地方，更是代表着阿克雅建筑事务所（意大利的建筑公司）要把自身和公司运营业务扩展到巴西的决

owner's architectural and family collection in the midst of an agricultural setting. As a place to preserve and reflect, this project explores the time element by setting itself unpretentiously on the site and behind the walls of an old barn. The partially restored old stone and brick walls tell the story of the past while two new contemporary timber structures behind it hints at the new. The scale is modest and fitting, without trying to stand out or be more than the nearby buildings. The simple yet bold gesture of the solid glue-laminated wall construction is a clever and effective approach by Hugh Strange Architects, based in London. Not only is this construction method good in stabilizing the internal environment for the archive, the architect efficiently meets the modest construction budget through high-quality prefabrication. The overarching corrugated roof unifies the structures into a single form. Thoughtful daylight openings punched through the roof provides the proper lighting for the archive. The finishing touch of this project balances the straightforwardness of the shell construction with a personalized balance of customized interiors. The craftsmanship of the displays and reuse of familiar materials reflects the unique background of the archive. The attitude of reusing nearby fallen wood for the construction of its hardwood flooring describes an awareness and respect for the resources that results in a personalized and comfortable interior. Together the result is a successful studio space that provides the ideal environment to reflect and discover the collection at hand. The disciplined design of this project earned a nomination for the 2015 European Prize for Contemporary Architecture, Mies van der Rohe Award and the RIBA National and Regional Award.

Archea studio in São Paulo is the new headquarters of Archea Associati. The notion of the studio, in this case, is not only a place to work, but also a representative space for Archea Associati, an Italian architecture firm, to expand their presence and operations in Brazil. This studio space is cre-

圣保罗阿克雅工作室，巴西
Archea Studio in São Paulo, Brazil

心。公司建立了工作室来支持其所需的新业务，即在莫伦比地区建造当代新型的住宅建筑。至于工作室的设计，新建的建筑物反映了阿克雅建筑事务所典型的特点。在阿克雅建筑事务所许多著名的设计作品中，比如，安东尼酒庄，都可以看到熟悉的深棕红色，这也可以在工作室的设计中看到。屋顶的金属框架和栏杆尤其如此，提醒着人们在其他项目中考顿钢的重要用途。从混凝土和暖色系马赛克露台地板的普遍使用，可以看出阿克雅对巴西环境的微妙适应。或许没有那么微妙，但是令人耳目一新并反映当地特色的是在面向后街的公司立面的涂鸦艺术。不管插画是否是建筑师特意选择的，但插画上使用的是适合公司外观的颜色，运用透视和超现实主义手法描述了一幅场景，就如同工作室内富有创造性的工作渗透到了室外。能感觉到，画匠们绘画时小心翼翼地描绘插画，没有污损窗框或者缠绕在拐角处。建筑设计以这种方式成为艺术的画布。工作室从外观上看是一个单一的体量，只是在前面和后面各有一个大窗户。从室内灰暗的茶色来看，这些会议室的景色衬托着周围涂鸦明亮的颜色。前门后是迎客的庭院，通过两层高的玻璃墙与工作室相连。又一次，鲜亮颜色的椅子和墙上的插画可以中和建筑物棕红色的色调。交通流线在室内与室外空间之间交织。两层高的建筑物室内设计很复杂，因为在每一个拐角处，不同的空间序列在亮度方面都要发生变化。地板的线性和天花板上的灯管成为一种特色，有助于指示方向。工作空间的大小合适，里面简单摆放着木桌，紧挨着玻璃墙，透过玻璃墙能看到庭院。以这种方式，阿克雅工作室展示着公司对空间感受和自身风格的理解，同时也在适应新的城市和气候环境。

在总部位于纽约布法罗的Davidson Rafailidis建筑工作室设计的"他，她和它"的项目中，可以看到实现空间设计和保证最大限度灵活

ated to support new work that the firm acquired, namely a new contemporary residential building in the city's Morumbi region. As far as the design of the studio office, the new building reflects characteristics typical of the Archea. The familiar deep brown red, as seen in many of their renowned work, such as Antinori Winery, is to be seen in this studio design. That is especially true with its metal trellis and railing on the rooftop that reminds one of their signature uses of their corten steel in other projects. The subtle adaptation to its Brazilian context could be seen with the predominant use of concrete, and the warm mosaics terrace flooring. Perhaps not so subtle but a refreshing local reflection is the graffiti art on the facade facing the back street. Deliberately chosen or not by the architect, the artwork uses colors that compliment the facade, depicting a scene with perspectives and surrealistic details as if the creativity work from inside the building oozes to the outside. The artwork somehow feels controlled, carefully not to deface the window frame or wrap itself around the corner. In this way, the architecture becomes the canvas for art. The otherwise monolithic exterior has only one large window on the front and one from the back. From the somber earth tones of the inside, the views of these meeting rooms frame the bright colors of surrounding graffiti. Behind the front door is a welcoming courtyard with an immediate reveal into the studio through the double-height glass facade. Once again, bright colorful chairs and artwork on the walls offset the brownish-red color scheme of the architecture. The circulation intertwines between the indoor and outdoor spaces. The interior of the two-story building feels intricate, as the sequence of spaces change in brightness with every turn. The linearity of the flooring and the ceiling light-tubes becomes a feature and helps guide orientation. The workspaces are modest, and are simply lined across wooden tables against the glass looking into the courtyard. Archea studio is the space that displays the firm's understanding of spatial sensibility

轻质棚屋,摄影师工作室,FT建筑师事务所
Light Sheds, Photographer's Studio by FT Architects

性的平衡。一对专业的、有创造力的夫妻把三个不同的工作室空间组合在一起,形成有棱角的屋顶结构。这对夫妻其中一位是画家,另一位是陶艺家、银器匠。这个项目迎合了需要各自工作空间的夫妻,他们各自有自己的原则。但同时,他们在共享第三个空间——温室的时候,也能在一起工作。每一个空间都有单独一个带角的屋顶,朝向相反的方向。画家的房间没有窗户,是为了大面积利用可用于绘画的墙壁,但是在倾斜屋顶的木梁上面有狭窄的天窗,让阳光均匀地照进荧光灯照明的白色工作室。陶艺工作室的窗户面朝正面,排成一排,获得更多的阳光。可移动的工作桌和各种照明设备可以给不同的陶艺作品提供开放的空间。一个角落里,木楼梯通向外形酷似桥的空间,这个空间盘旋在工作室的上方。这个小小的空间里,两边设有桌子,还有制作珠宝时所必需的吊灯。工作室的一头,从大窗户望去,可以看见外面自然的景色,阳光也可以透过窗户照进来。在另一头,从窗户可以看到下方相邻温室的内部景色。温室的外形是长方形的,四周装满了聚碳酸酯板来控制适宜植物生长的温度,这里也是夫妻俩逃离尘世、放松休闲的第三个场所。在一层,隔墙把这三个空间隔开,使之相互独立。把隔墙打开时,隔墙就好像在开阔的一层空间上漂浮,从而形成大面积的灵活空间。采光口之间跨度之大,令人印象深刻,这是通过一段段的高墙实现的,具有独创性。这些高墙会产生刚度,和结构桁架的作用一样,创造出一种流动的效果,好似摆脱了重力的束缚。这座面积为140m²的建筑尽管体现了夫妻二人各自独特的喜好和生活方式,但却是不同工作室空间成功结合的产物。

位于日本关东的一家小型摄影师工作室在设计时运用照片拍摄时的大量灯光,探索了创造艺术工作环境的方式。几何形状的木框架可

and their own style, while adapting to the new urban and climatic context.

The balance between design-specific spaces and maximum flexibility could be seen in the project "He, She & It" by architecture studio Davidson Rafailidis, based in Bufallo New York. Three distinct studio spaces come together into one composed structure of angular roofs for a professionally creative couple; one is a painter and the other ceramist/silversmith. The project caters to the couple, each needing their specific work spaces for their own disciplines but at the same time, it allows them to work alongside each other while sharing a third space, a greenhouse. Each space has a single angular roof that pitches towards another direction. The painter's space has no windows in order to utilize the large workable wall space, but a strip of skylights above wooden beams of the sloped roof provide daylight into otherwise an evenly florescent-lit white studio. The ceramic workspace has more daylight as a band of windows lines the facade. Movable worktables and various lighting serve the various ceramic work of the open space. On one corner, a wooden staircase leads to a bridge-like space that hovers above the workshop below. This intimate space is lined with tables on both sides and hanging lamps for fine jewelry-making work. On the one end, a large window welcomes the exterior view of nature and daylight to pour in, and on the other, a window peaks into the interior of the adjacent greenhouse below. The greenhouse is a rectangular space cladded on all sides with polycarbonate panels to control the climate for plants, but it also acts as a third place where the couple could refuge and relax with one another. On the ground floor, partition-walls borders separate the three individual spaces. When open, the walls above seem to float over the open ground floor, creating a large flexible space. The large spans of the openings are impressive, and are cleverly achieved through the high wall segments that produce stiffness and work as structural trusses to create the

素描、绘画、雕刻小工作室，Christian Tonko
A Small Studio for Drawing, Painting and Sculpture by Christian Tonko

以支撑极具雕塑感的屋顶结构，而不需要水平方向的托梁。拆除水平方向的结构为拍摄照片提供了合适的高度。FT建筑师事务所成功地把传统的木材技巧和现代的建筑方法结合起来。半透明聚碳酸酯板的波形表皮显示了框架结构的工艺水平。

位于布雷根茨的小型管状工作室是Christian Tonko设计的，融入了周围奥地利的景色中。有趣的是，建筑物的形状似乎来源于其本身想要"框住"当地的景色。工作室的目的是激励人们去素描、去绘画、去雕刻。其结构似乎融入了周围风景，用两个楼层迎合了当地的地形地貌。建筑上层是用来素描和画水彩画的工作空间，而下层是用来绘画和雕刻的场所。

工作室空间的设计会因需求、风格和功能的不同而存在差异，但是最根本的意图是建造一个安全的地方，供人沉思、学习并进行富有想象力的生产。所以无论是摄影、建筑、绘画、陶瓷或是珠宝，工作室就是避难所，在这里，人们可以不分昼夜地工作，永不停歇。这是一个令人备受鼓舞的场所，让人得到激励；这是一个富有创造力的地方，让人从中获得创造力。建造这些空间时，应该精确地满足富有创造力的专业人员纪律上的需要，允许他们有空间和灵活性去测试、研究和体验他们的艺术。

floating effect, defying gravity. The 140 square meter building is a successful marriage of distinct studio spaces while reflecting the particular likings and lifestyles of the clients. The design for a small photographer's studio in Kanto, Japan, explored ways of creating an artistic working environment with lots of ambient light for photo shoots. The geometric timber frame supported a sculptural roof structure without horizontal joists. Getting rid of the horizontal members allows the clear height desired for photo shoots. FT Architects was able to merged traditional timber techniques with modern building methods. The translucent polycarbonate corrugated skin reveals the craftsmanship of the framing work.

The small tube-like studio that Christian Tonko designed in Bregenz frames its views to the surrounding Austrian landscape. Interestingly, the form of the building seems to be derived from the views it wants to "frame". The studio aim to inspire one to draw, paint and make sculpture. The structure appears to hug the landscape. It accommodates the topography with two separate levels. The upper level was designed to be workplace for sketches and watercolors, while the lower portion was designed for paintings and sculptures.

The designs of studio spaces can differ in needs, style, and functions, but the underlying intention is to make a safe place to reflect, study, and imaginatively produce one's craft. So whether it's photography, architecture, painting, ceramics, or jewelry, the studio is a sanctuary where one could work away and be in the flow. It is an inspiring place to be inspired, and a creative place to be creative. The architecture of such space should address the precise disciplinary need of the creative professional, but allow the room and flexibility for one to test, research and spontaneous experiment with one's art. Andrew Tang

轻质棚屋，摄影师工作室
FT Architects

实用主义 + 象征主义

本案客户是一名摄影师，他的老房子位于关东南部的丘陵地区，花园里有个轻质棚屋，4.5m×7.2m 单间，是摄影师的工作室，具备所有摄影工作室所需的基本功能。起初，建筑师想法很简单，建一个简单实用的棚子。但是客户预算有限，为了在预算范围内建一个宽敞的房间，建筑师选择木框架人字形屋顶。不过屋顶通常需要横向托梁，这样的话就会降低拍摄照片所需的净高。

根据简单的几何原理，建筑师设计了多面的、非对称的屋顶形式。折叠线上的三个脊梁支撑着屋顶结构，不需要任何可能影响房间高度的水平构件。建筑师之所以选择了原木作为脊梁，是因为其能适应木椽多角度组装的复杂性。由于聚碳酸酯面板是半透明的，舞台背景的背面木材框架清晰可见。

和艺术家工作室一样，摄影工作室也需要有充足阳光的环境。但是阳光直射会造成对比度过强，导致拍照不自然。45°天窗与被两个不同角度漫射光笼罩的窗台相结合，加上宽阔的窗户朝向郁郁葱葱的花园，室内虽然封闭着，但却总是和室外一样明亮。建筑兼具内外双重性，模糊了传统建筑的定义。

自FT建筑师事务所第一个项目开始就与日本木材制造商"c-office"合作，该公司擅长采用实用结构解决方案，这一方案一直是事务所所有项目的核心。FT建筑师事务所也在不断探索各种方法来继承日本传统木制品的精粹，并将之应用于现代建筑中。在这个项目中，原木梁成为解决方法以满足设计任务书的要求，同时它还具有象征意味，是最古老的建筑材料之一。建筑师在这里使用木材制作人造场地，为摄影师提供了可以在"自然"光下拍摄物体的环境，从而检验了木材这种原始材料的现代应用。

Light Sheds, Photographer's Studio

Pragmatism + Symbolism

The light sheds is a photographer's studio that stands in the garden of his existing house, located in the hilly region of southern Kanto. The space is solely composed of essential functions required for a photographic studio and the result is one single room of 4.5m by 7.2m. From the start, we had a simple concept of creating a simple, pragmatic sheds.
In order to form a large volume out of the limited budget available, a timber frame, gable roof structure was selected. However, a gable roof would typically entail horizontal joists, which would reduce the clear height required for photo shoots. Following simple geometric rules, the roof was distorted into a multi-facetted, asymmetric form. Three ridge beams at the folding lines support the roof structure, negating the need for any horizontal members that may compromise the height of the room. Logs were selected for the ridge beams, for their ability to accommodate the complex assembly of timber rafters at various angles. Translucent polycarbonate panels reveal the "back of the stage set" timber framework of the walls.

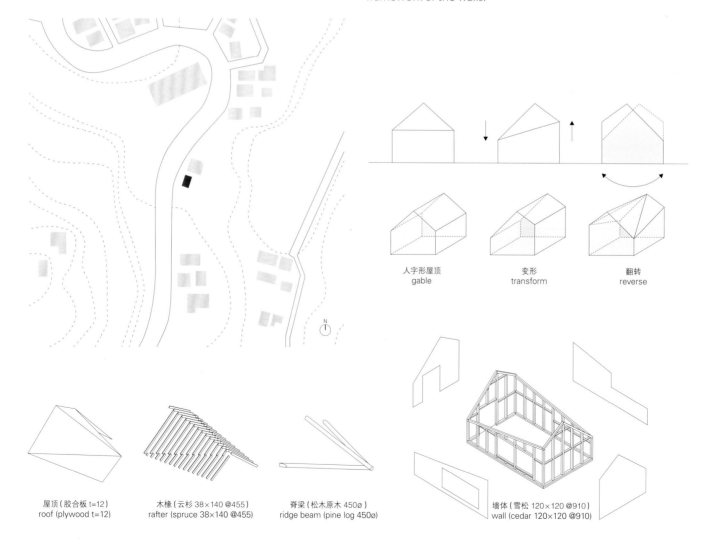

人字形屋顶 gable　　变形 transform　　翻转 reverse

屋顶（胶合板 t=12） roof (plywood t=12)
木椽（云杉 38×140 @455） rafter (spruce 38×140 @455)
脊梁（松木原木 450ø） ridge beam (pine log 450ø)
墙体（雪松 120×120 @910） wall (cedar 120×120 @910)

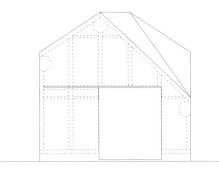
南立面 south elevation

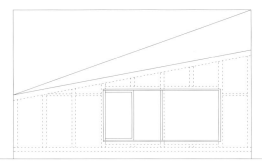
东立面 east elevation

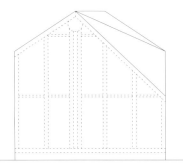
北立面 north elevation

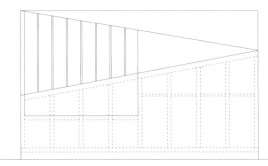
西立面 west elevation

项目名称：Light Sheds, Photographer's Studio / 地点：Kanagawa, Japan
建筑师：Katsuya Fukushima, Hiroko Tominaga
合作者：Shu-tada Structural Consultant / 用途：photo studio
用地面积：422.41m² / 建筑面积：33.12m² / 有效楼层面积：33.12m²
结构：wood / 设计时间：2013 / 施工时间：2014.4—2014.7
摄影师：©Shigeo Ogawa (courtesy of the architect)

Like artists' ateliers, photographic studios require ample ambient light. Direct sunlight creates overly strong contrasts that would result in unnatural portraits. A combination of skylights at 45 degrees and a clerestory let in diffused light from two different angles from above. Together with the wide windows on the side facing onto the lush garden, the interior, although enclosed, is always as bright as the external environment. It has the duality of being both outside and in, blurring the conventional architectural boundaries.

Ever since the practice's first project, c-office, an office for a Japanese timber manufacturer, application of pragmatic structural solutions has been the core to our approach in all projects. We have also been continually exploring ways to inherit and translate the purity of traditional Japanese timber compositions into modern construction. Here, while the log beams were employed as pragmatic means to fulfil the brief, they also bring along symbolic associations of being one of the oldest building material. In this project, we were able to test the modern application of this primitive material by using it as a meaningful device to create a man-made enclosure where photographic objects can be exposed in "natural" light.

FT Architects

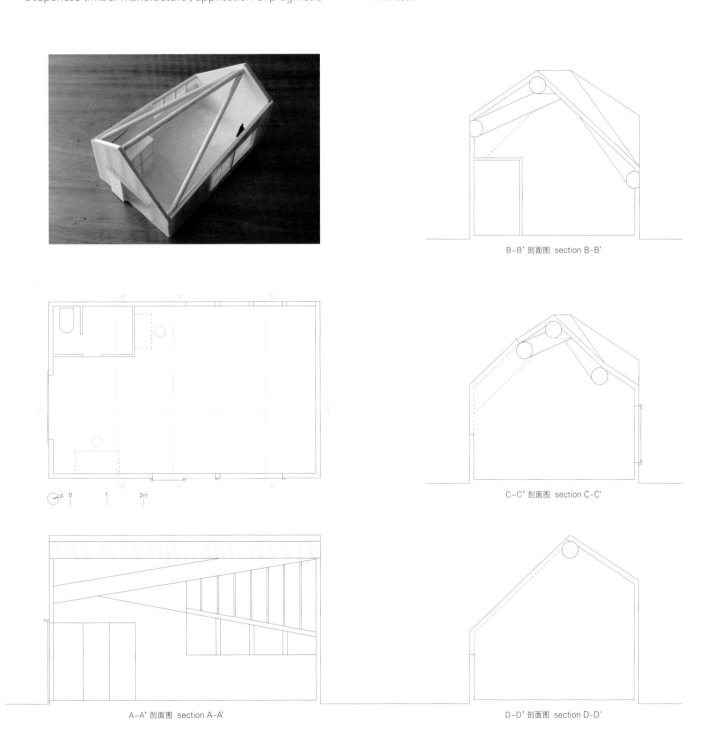

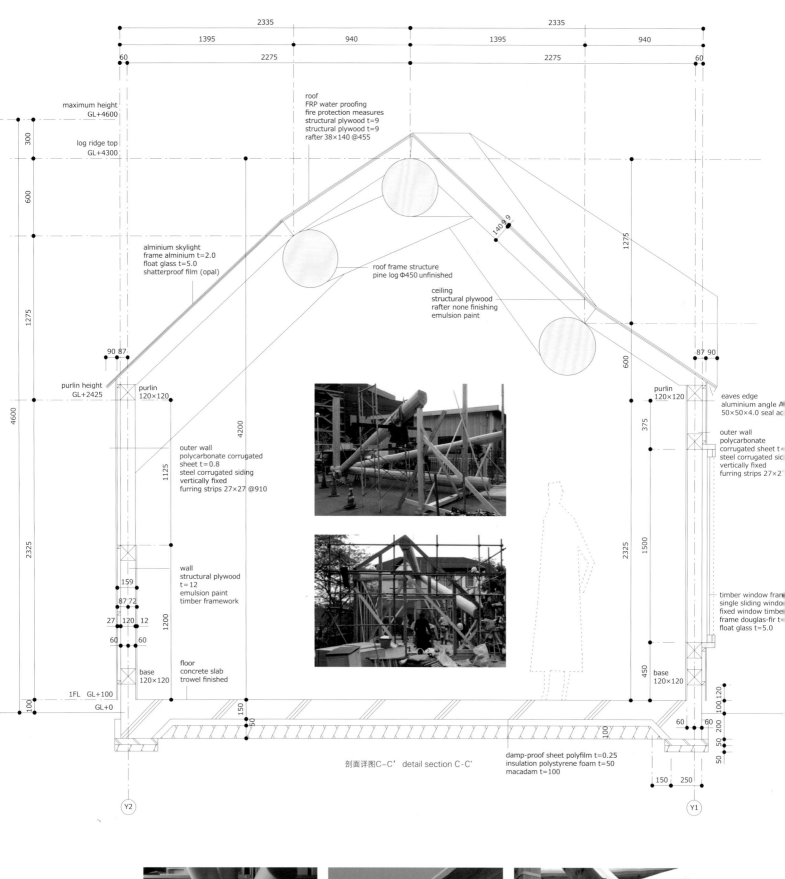

剖面详图C-C' detail section C-C'

他，她和它
Davidson Rafailidis

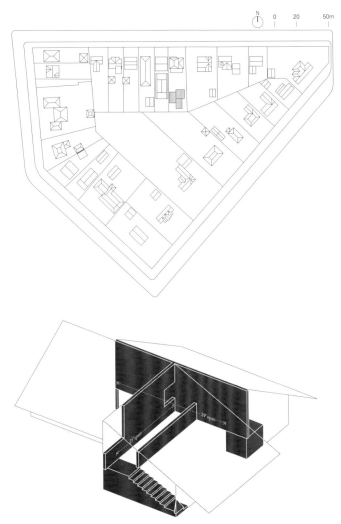

　"他,她和它"是三个不同建筑的集合体,它将满足不同空间需求的三座建筑拼贴成单一结构。135m² 的建筑里有画室、陶艺／银器工作室和温室。每个空间都呈现出完全不同于其他的空间氛围。每个截然不同的空间氛围不但体现了它们各自的用途,也反映出客户的个人偏好。

　他是个画家,他的工作室是个白色的盒子,没有窗户。唯一的照明来自顶光,提供均匀的间接光、自然光,将墙面的绘画面积最大化。

　她是一位陶艺师和银器匠。她的工作空间分为两个:一个专门用于陶器制作,凌乱、潮湿;另一个用于制作精致的珠宝首饰。她的空间窗户巨大,视野宽阔,光线变化显著,有的地方光线暗淡,而工作台光线明亮。工作室墙面用柔软光滑的枫木做装饰,保留了其浓郁的原木味道。

　　它（他们）有春天的幼苗和冬天的植物——客户的愿望很简单：光照最大化，全年不结冰。半透明的聚碳酸酯外壳为其他两个工作空间提供了几乎相当于室外的一个空间地带，但又不会相互直接看到彼此。

　　三个空间连为一体，形成了由三个单坡屋顶棚屋组成的建筑群。在三座不同的建筑相互连接的地方，2m 以下的墙体都被拆除了；而上面所剩的顶墙部分就像建筑桁架一样自由地横跨半空，使得这一构造看起来像是悬停在空旷的一层之上。"他，她和它"的三个结构看起来相互独立，建筑材料用量也明显不同，但在结构和室内气候上却相互依存。折叠推拉门让使用者可以将其分割为三个房间，也可以把整个平面全部打通——随时可以按照需求重新分割布局。

　　气候调节策略强调动态的空间体验和空间的不断重新定义。每个空间都提供了不同的气候屏障。必须通过带保温的室内滑动折叠门和外部可开启的洞口来调节空间温度，使其适应不同的天气状况。不同于常见的气候控制做法（尽可能将室内空间与室外空间密封隔绝，冬天使用机械设施在室内制造出人为的夏天），"他，她和它"的建筑师根据季节调整空间而不是采用机械手段。这两种不同环境控制手段间的差异可以用机动船和帆船进行类比。机动船依靠机器达到恒定的速度和方向，而帆船则通过天气、风向、航行几何学和水手的积极参与之间的亲密关系不断地改变帆船／航行的整体几何结构，达到同样的目的。例如，在寒冷晴朗的冬天，滑动折叠隔墙完全打开，让从温室获得的太阳得热为整栋建筑贡献热量。在阴冷的冬日和夜晚，带保温的隔墙则需要关闭以缩小整体需要供热的体积。在夏天，打开温室连续的屋脊通风孔可以将温室变成一个太阳能烟囱，即使在没有一丝风的炎炎夏日也能在整栋建筑中制出源源不断的气流。

单坡体量："他，她和它"	平面示意图："他，她和它"，室内布局变化	单独平面图	他分开，她和它连在一起	她分开，他和它连在一起	它分开，她和他连在一起	他和她分开，共用它
mono-pitched volumes: He, She & It	plan diagram: He, She & It, with internal configuration variations	separated plan	He separated, She and It connected	She separated, He and It connected	It separated, She and He connected	He and She separated, It shared by both

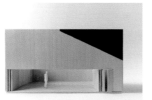  

他 He　　她 She　　它 It

共享平面图
shared plan

项目名称：He, She & It
地点：Buffalo, NY, USA
建筑师：Davidson Rafailidis
合伙人负责人：Stephanie Davidson, Georg Rafailidis
助理：Alex Marchuk, Jia Ma
模型制作：Matt Meyers
结构工程师：John Banaszak
植物与地面：Matt Dore
用途：Artist studio
建筑面积：135m²
竣工时间：2015
摄影师：©Florian Holzherr (courtesy of the architect)

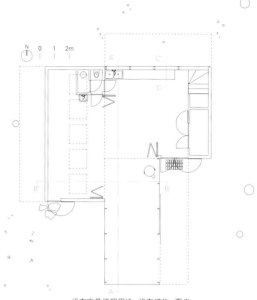

没有家具指明用途，没有植物，夏天
no furnishing to indicate use, no vegetation, summer

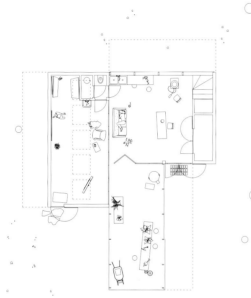

家具指明用途，没有植物，冬天
furnishing to indicate use, no vegetation, winter

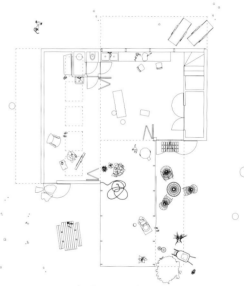

家具指明用途，没有植物，夏天
furnishing to indicate use, no vegetation, summer

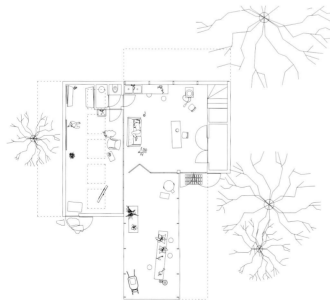

植物和家具指明用途，冬天
vegetation and furnishing to indicate use, winter

植物和家具指明用途，夏天
vegetation and furnishing to indicate use, summer

这座以成本价较低的材料和建筑方法打造的建筑物，其空间安排在丰富的空间和组织尺度方面提供了一套复杂的经验。工作空间的室内引人入胜，成为客户逃避外部世界的休息港湾。入口门在工作室的后面，将走廊尽可能延伸，使其远离临街的居住区，强调工作空间与前面常规的住宅建筑的本质区别。三个单坡屋顶有着巨大的悬挑，使雨水远离建筑物，滋养三个雨水花园，进一步将建筑与前面的住宅隔离开来。

He，She & It

He, She & It is a collection of three distinct buildings for three different spatial needs, collaged into a single structure. The 135 sqm building houses work spaces for a painter, a ceramist/silversmith, and a greenhouse. Each space offers an atmosphere which differs radically from the others. The distinct atmospheres of the spaces reflect not only their respective uses, but also, the predilections of the clients.

He is a painter. His studio is a white box. There are no windows in his work space; it is exclusively top-lit, offering even and indirect, natural light, and maximizing the wall surface area for painting.

She is a ceramist and a silversmith. Her work space has dedicated areas for both messy, wet ceramic work and delicate jewelry-making. Her space offers large windows with generous views and dramatic lighting, ranging from dimly-lit areas to very bright desk areas. Her studio is lined entirely with soft, soaped, maple, preserving its intense, raw wood smell.

It (they) consists of seedlings in spring and plants in winter – clients with a very simple wish for maximum light and year-round above-freezing temperatures. The polycarbonate shell is translucent, offering a zone of almost-outdoor space to the two other work spaces, without any direct views.

The spaces are grouped to form a cluster of three mono-pitched sheds. At the surfaces where these distinct sheds connect, the walls are completely removed up to a height of 2m. The remaining ridge wall segments above act as structural trusses to span the openings freely, making the structure seem to hover over the open ground floor. Whereas He, She & It appear as independent, materially distinct volumes, structurally and climatically they depend on one another. Folding-

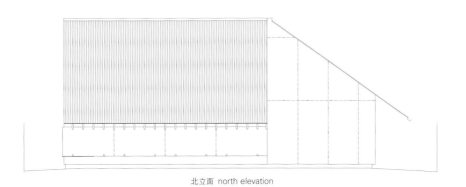

北立面 north elevation

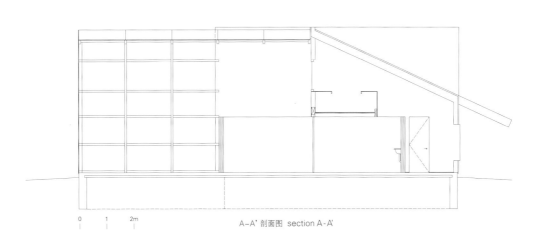

0 1 2m

A-A' 剖面图 section A-A'

sliding doors enable the users to either divide the space into three rooms, or open the plan entirely – it is in a constant state of redefinition.

The climate conditioning strategy underscores the dynamic spatial experience and the constant redefinition of space. Each space offers a different climatic barrier. The insulated interior sliding-folding doors and exterior operable openings have to be used to adjust the space to different weather conditions. Contrary to common climate control practices that seal the interior space as much as possible from the exterior and use mechanical services to create an artificial indoor climate in summer as in winter, He, She & It adapts spatially to the seasons instead of mechanically. This difference is analogous to a sailboat versus a motorized boat. Whereas the motorized boat achieves a constant speed and direction through a machine, the sail boat achieves the same through an intimate relationship between weather, wind, sail geometry and the sailor's active involvement, changing the overall geometry of the sailboat/sails constantly. In cold and sunny winter days, for example, the sliding folding partition walls are opened up to let the solar gain from the greenhouse contribute heat to the whole building. On cloudy cold winter

days and winter nights, the insulated partition walls need to be closed to shrink the overall heated volume. In summer, the continuous ridge vent of the greenhouse is opened-up, transforming the greenhouse into a solar chimney that creates constant draft throughout the building even on stagnant, hot days.

Build with modest, low cost materials and construction methods, the spatial arrangement offers a complex set of experiences that are rich at the spatial and textural scales; the interior world of the workspaces draws the users in and provides them with a retreat from the outside world. Entry doors are located at the back of the studio, extending the passage as far as possible away from the existing residence facing the street, and emphasizing the distinct nature of the workspace from the residential routine of the front house. The three mono-pitched roofs have large overhangs to shed storm water far away from the building and nurture three rain gardens that act also as visual screens, isolating the building even more from the front house.

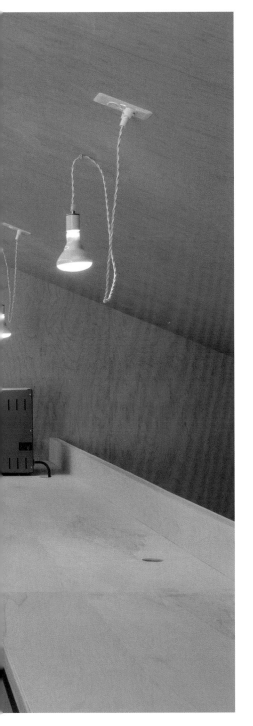

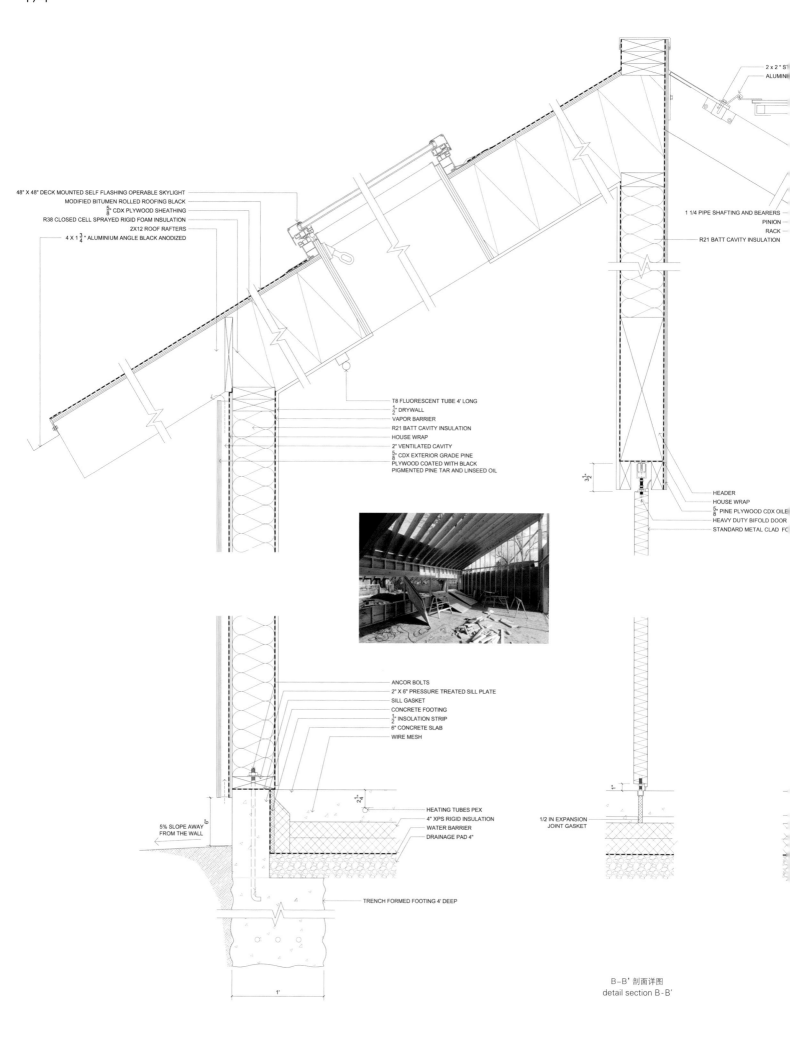

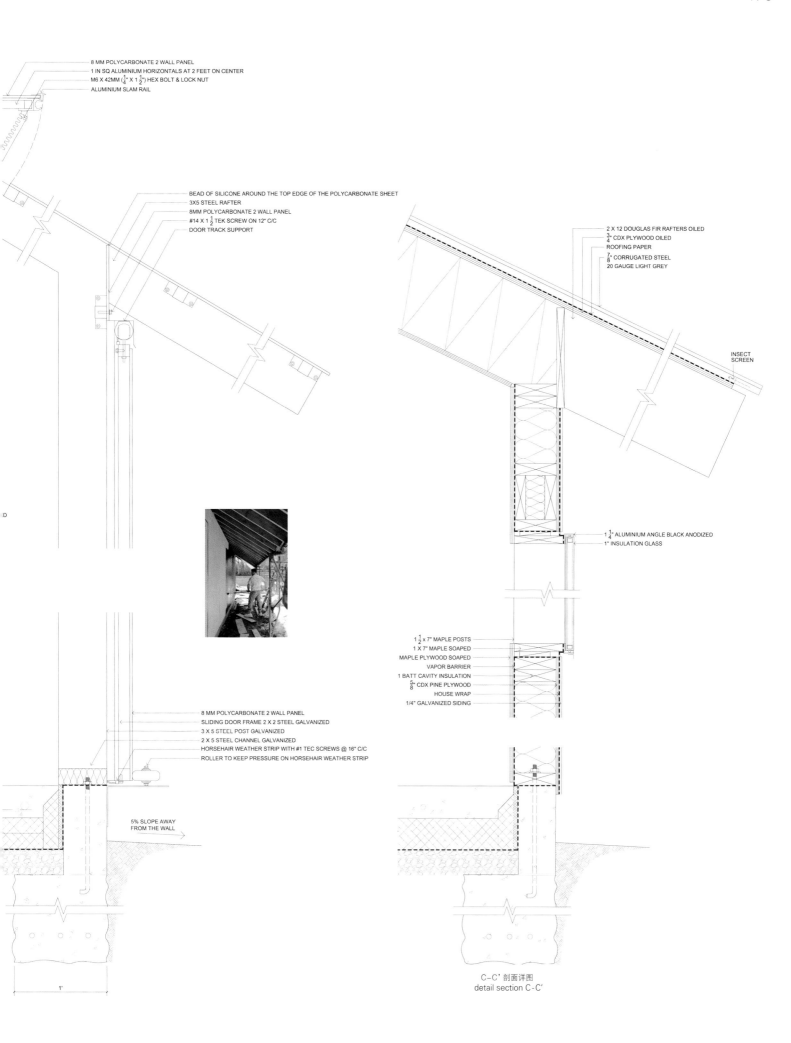

C-C' 剖面详图
detail section C-C'

这个用来进行素描、绘画和雕刻的小型工作室是一座视野通透的建筑，两个方向都可以看到周围的景色。在建筑的东南面，大量的阳光可以透过倾斜的窗玻璃照进来。为了阻断阳光的照射以及调节光和气候条件，建筑师设计了外部隔板。建筑西北面安装了框架系统，可以使青铜雕像悬在玻璃前面，正对着工作的艺术家的视线。在那个位置，青铜雕像保有天然形成的铜绿，同时能使艺术家进行反思，专注于自己的工作。

这栋建筑的特点是有独立的两层，各有不同的功能。上层是办公空间，大多数的素描和小型的水彩画都是在这里完成的；而下层是创作中型的油画以及小型雕塑的地方。这个项目半工业的特点源于对单坡屋顶工厂外形的参考。此处的外形指的是被简化为最简单形式的一个有天窗的盒子。

原材料的使用更加凸显了工厂的特点。正面的镶板是由耐候钢制成的，而内表面是由粗糙混凝土、粗钢和未处理的橡木制成的。

项目最终的布局由现场条件决定，场地的边界合适，对附近的居民楼干扰程度最小，符合高度限制要求，适应山坡情况，符合对楼板面积和楼底高度的特定要求。从基本概念角度来说，设计受到一种古代光学装置——投影描绘器——的启发。一方面，从字面上看，它是一种明亮的仓室，可以形成良好的光照条件，可以调节到想要的光照水平。同时，这个工作室自己就是一个光学调节装置，这一点和投影描绘器作为绘图辅助工具的原始功能相似。

A Small Studio for Drawing, Painting and Sculpture

This small studio for drawing, painting and sculpture acts as a visual device itself by bidirectionally framing its surroundings. To the southeast a great amount of daylight enters through tilted glazing. To block direct sun if desired and to enable the modulation of light and climatic conditions exterior screens are deployed. To the northwest a system of frames is installed which enable bronze sculptures to be suspended in front of the glass and in direct sight of the working artist. In that spot the bronze sculptures receive their natural patina while being staged as a motive of reflection and confrontation for the artist.

The building features two separate levels which serve different functions. The upper level is designed to be a workplace where most of the sketches and small water colors are done while on the lower level medium sized canvases and small sculptures will be produced.

素描、绘画、雕刻小工作室
Christian Tonko

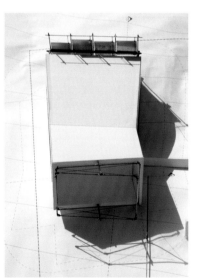

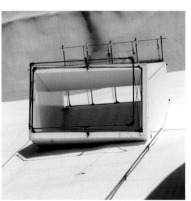

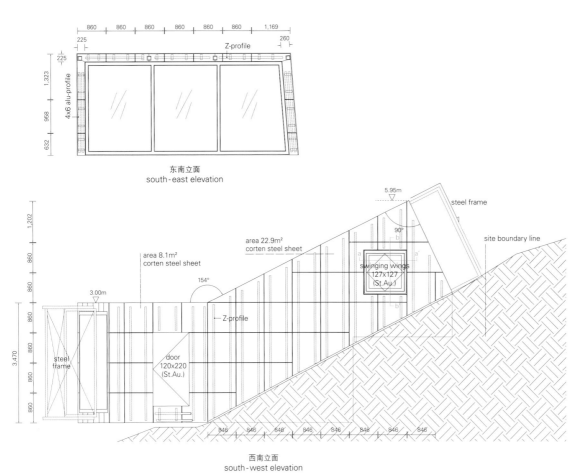

东南立面
south-east elevation

西南立面
south-west elevation

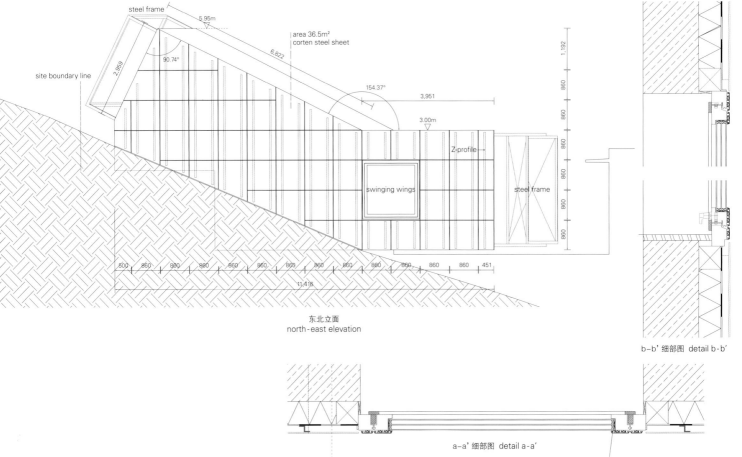

东北立面
north-east elevation

a-a' 细部图 detail a-a'

b-b' 细部图 detail b-b'

The semi-industrial character of the project stems from the reference to the typology of the shed roof factory. Here this typology is being reduced to its simplest case – a single box with a single skylight.

The use of raw and untreated materials contributes to the character of a workshop. The facade panels are made from weathering steel while the interior surfaces are made from raw concrete, raw steel and untreated oak.

The final configuration of the project remains determined by site conditions as it is fitted into the plot boundaries while negotiating a desired minimal disturbance of the neighbouring residential building, height restriction and the hillside situation with certain requirements of floor area and ceiling height. On an underlying conceptual level the design is inspired by an ancient optical device – the camera lucida. On the one hand it is very literally a bright chamber – constructed to achieve good light conditions which can be modulated to desired levels. At the same time the studio itself acts as an optical framing device similar to the original function of the camera lucida as a drawing aid.

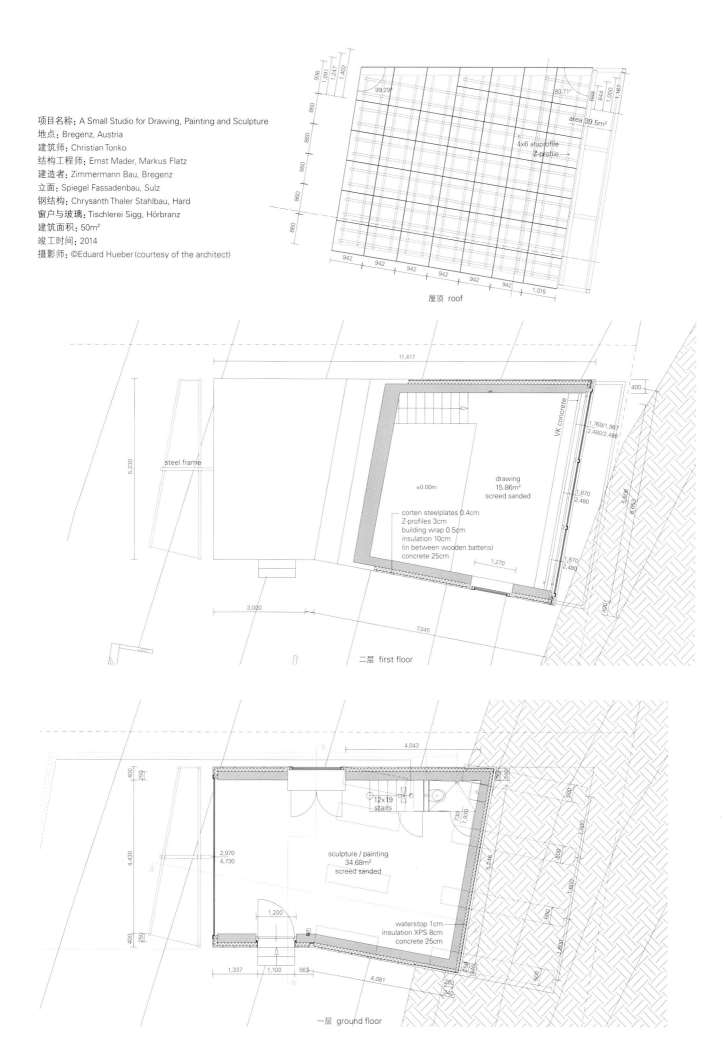

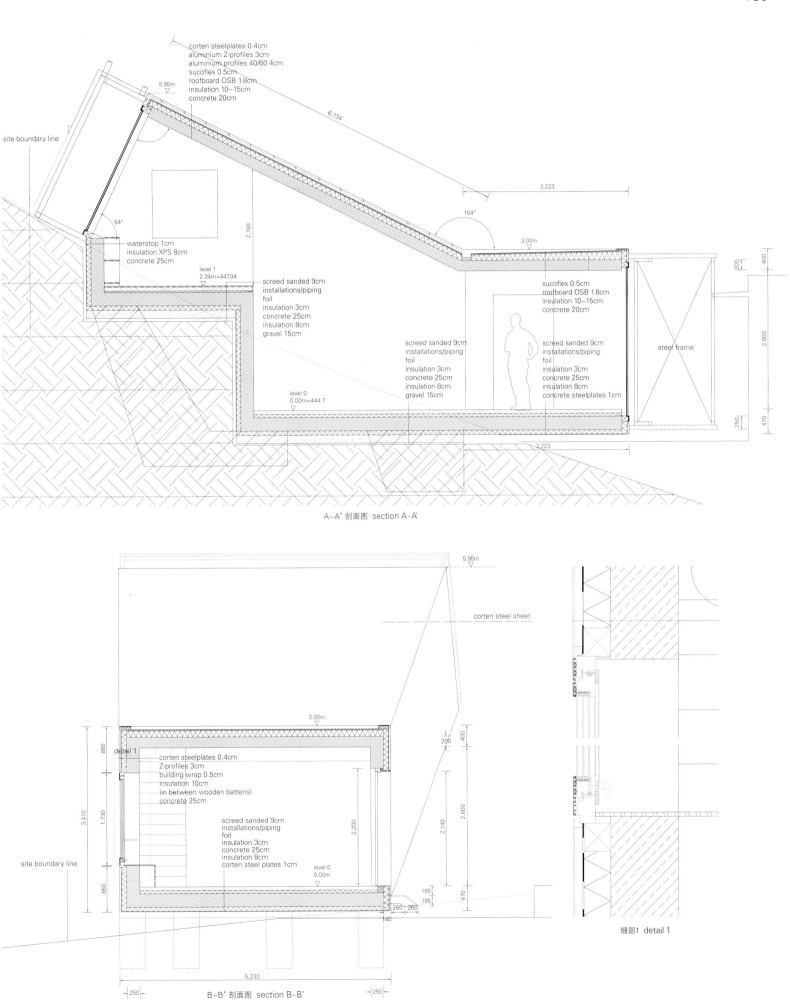

位于萨默塞特农场环境中的新建筑档案馆和工作室坐落在一个农业耕种山谷下，周围的建筑物还有谷仓、木材商店和一些小棚子，这些棚子可以追溯到19世纪，从20世纪70年代到最近，完全变成了牛棚。一个古老的砖石牲口棚已经破烂不堪，摇摇欲坠，墙体和屋顶已被拆掉，建筑师对其余的墙体进行了仔细修复和加固。建筑师在这些建筑之间增加了两个木结构，新建屋顶一直延伸至建筑整个长度，提供了一个带顶的入口。两个木结构的体量完全相同，但通过门窗布局和装修进行了巧妙区分。北边的建筑是一个办公空间，其突出特点是采用大落地窗，能将外面树木繁茂的山谷景色尽收眼底。与其形成对比的是，南边的建筑是一个建筑存档和展览空间，其显著特点是主要通过大量的顶灯进行顶部照明。建筑的后面有个不大的外部空间，由保留下来的墙体和山谷的斜面封闭，并通过一部宽大的室外楼梯通向周围的林地。

新建筑的外壳是一层结实的交叉层压板，没有设置保温层、外覆层和内衬层。木板的厚度从300mm到420mm不等，可以同时提供隔热和蓄热功能，为建筑图纸的储藏创造相对稳定的温度和湿度。坐落在粗糙浇筑混凝土地基上的简单木结构像周围的谷仓一样由异型水泥屋顶加以保护。木结构和木覆层之间的通风空间可以防止夏天过热。

与工程用云杉结构不同，两个大房间的地板使用了从周围林地砍伐并在附近的木材商店变干的硬木材。办公室使用粗锯的雪松地板，档案室使用砂纸打磨然后上油的白蜡木和榉木，将两个不同特点的空间区分开来。特别委托制作的展板和窗帘与标准的包装毯和定制的工艺品融为一体。该项目的工程预算是用25万英镑提供120m² 的使用面积。项目团队将简单的建筑技术、预制组件、廉价的工业材料和高品质的当地硬木结合起来，并通过显而易见的经济手段建造了这座建筑。它质朴地坐落在农场中，却提供了宽敞的空间和奢华的室内设计。

Drawing Archive and Studio, Somerset

Located within the context of a working Somerset farmyard the new Architecture Archive and Studio sits at the foot of an agricultural valley within a mix of buildings that includes barns, wood stores and sheds that date from the 19th Century, through the 1970's, to a recently completed cowshed. The dilapidated walls and roof of an old stone and brick barn have been removed and the remaining walls carefully stabilized and repaired. Within these, two timber structures have been inserted, with a single new roof extending the full length of the building to provide a covered entrance. Identical in volume, the two timber volumes are subtly differentiated through their fenestration and fit-out. The north building provides an office space and is characterised by large French windows that offer views to the wooded valley outside. In contrast the south building houses a drawing archive and

萨默塞特建筑档案馆和工作室
Hugh Strange Architects

项目名称：Architecture Archive and Studio, Somerset
地点：Shatwell Farm, Yarlington, Somerset
建筑师：Hugh Strange Architects
结构工程师：Price & Myers / 木结构工程：Eurban Ltd
展板：Jude Dennis, Lars Wagner / 客户：Niall Hobhouse
用途：Office / 施工：Paul Rawson / 竣工时间：2014.1
摄影师：©David Grandorge (courtesy of the architect)

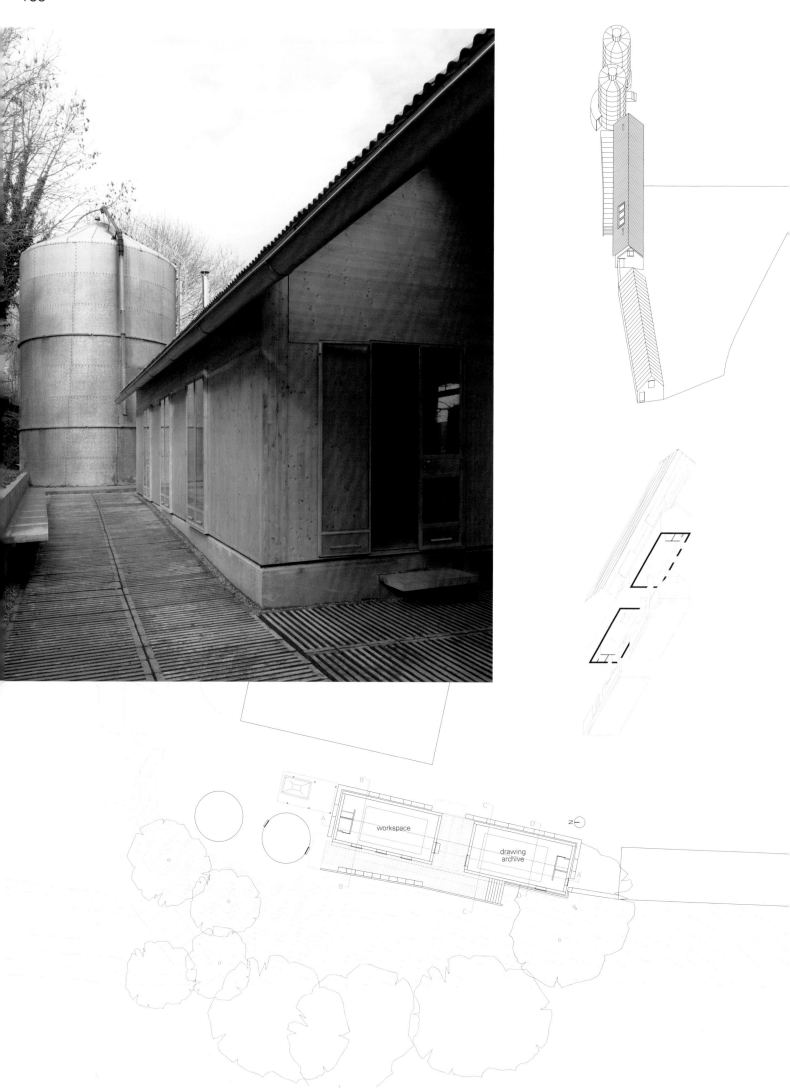

display space and is predominantly top lit by a generous roof light. To the rear of the building a modest external space is enclosed by the retaining wall to the sloping side of the valley, with a generous external stair giving access up to the surrounding woodlands.

The new building shell is constructed of a single layer of solid cross-laminated timber without insulation, external wall cladding or internal lining. The wood panels range from 300mm to 420mm in thickness and simultaneously provide insulation and thermal mass, creating stability of temperature and relative humidity for the drawing's storage. Sitting on a rough in-situ cast concrete base the simple timber forms are protected by a profiled cement roof similar to those of the surrounding barns. The vented space between this and the timber cladding prevents overheating during hot summer months.

In contrast to the engineered Spruce construction, the floors of the two large rooms are fit out with hardwood "mats" using timber felled from the surrounding woodlands and dried in the neighbouring wood store. Rough-sawn Cedar floorboards to the office and smooth-sanded and oiled Ash and Beech to the archive distinguish the different characters of the spaces. Specially commissioned display panels and curtains marry standard packing blankets with bespoke craftsmanship.

With a construction budget of GBP 250,000 and providing 120m² of accommodation, the project mixes simple construction techniques, pre-fabrication and cheap industrial materials with high-quality locally sourced hardwood to create a building that, through an evident economy of means, sits unpretentiously in its farmyard environment, while providing generous and luxurious interiors.

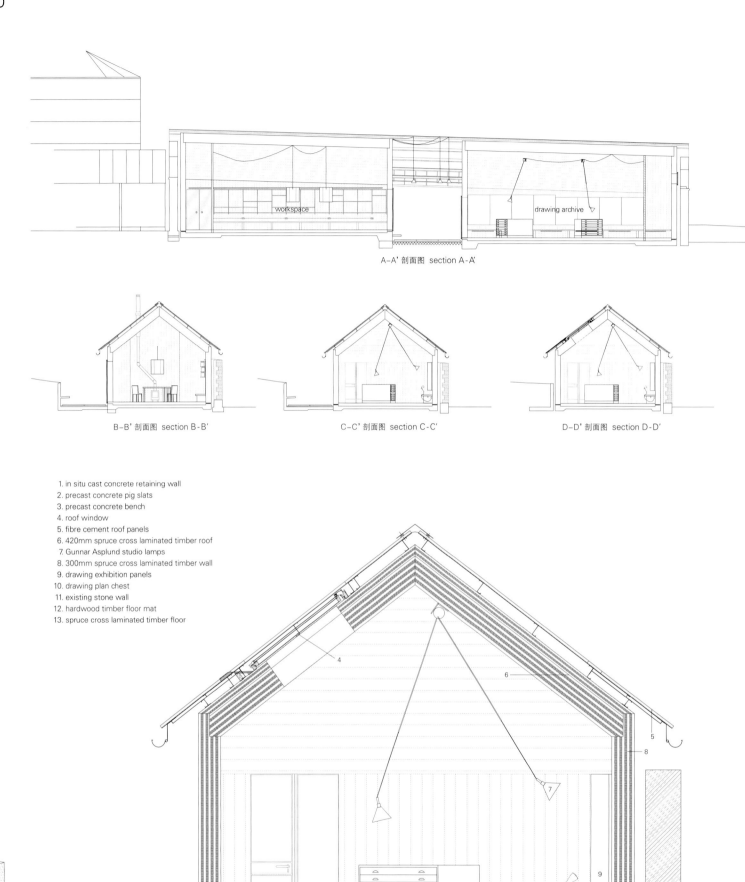

A-A' 剖面图 section A-A'

B-B' 剖面图 section B-B'

C-C' 剖面图 section C-C'

D-D' 剖面图 section D-D'

1. in situ cast concrete retaining wall
2. precast concrete pig slats
3. precast concrete bench
4. roof window
5. fibre cement roof panels
6. 420mm spruce cross laminated timber roof
7. Gunnar Asplund studio lamps
8. 300mm spruce cross laminated timber wall
9. drawing exhibition panels
10. drawing plan chest
11. existing stone wall
12. hardwood timber floor mat
13. spruce cross laminated timber floor

C-C' 剖面详图 detail section C-C'

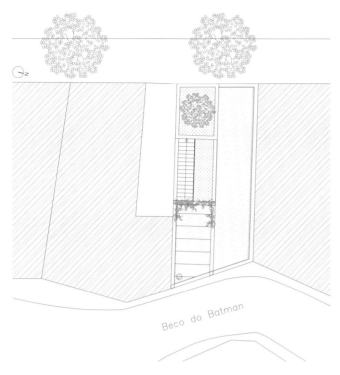

Rua Harmonia

Beco do Batman

圣保罗阿克雅工作室

Archea Associati

  2012年8月，阿克雅建筑事务所在拉丁美洲最大的城市圣保罗开设了分支工作室。圣保罗位于巴西的东南部，被视为南美大陆的商业中心，在当今发展阶段是巴西最有发展前景的城市之一。这个城市有2000万居民，不仅提供了很多房地产建造的发展机会，经常有建筑设计机会，还有运动、文化、社会活动机会以及移动电信领域的创新发展机遇。

  因此，阿克雅建筑事务所在Morumbi居民区附近开设了新的工作室，Morumbi距离圣保罗市中心约有9000m到15000m远。这是阿克雅在巴西的第一个建筑项目。

  这栋三层的建筑在周围的开发区域里很显眼，这是因为它独特的搭配：浓厚的深棕色与混凝土的独特质地浑然一体。形状像调色板的涂鸦艺术品直接放在人行道上，开出了一条窄路。建筑的另一面，窗户、入口、墙以及墙上的铁栅栏简单地排列着。

  从入口进去就是光线柔和的大厅，由考顿钢覆盖的大会议室就在此处。

Archea Studio in São Paulo

In August 2012, Archea Associati opened a branch studio in São Paulo, the biggest city of Latin America. Located in the southeast of Brazil, the city is considered the business center of the South American continent and one of the most promising areas in the current phase of the development of Brazil. With more than 20 million inhabitants, the city offers many opportunities not only in the sectors of real estate and constructions, and consequently that of architecture, but also

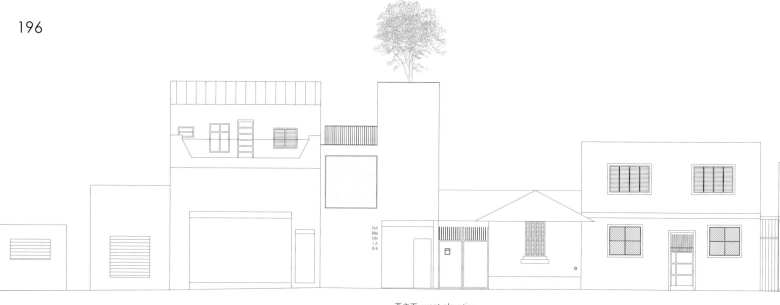

西立面 west elevation

东立面 east elevation

夹层 mezzanine

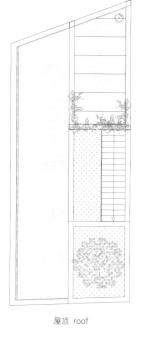

屋顶 roof

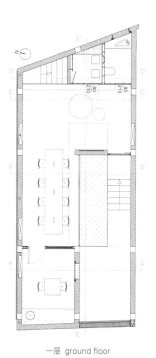

一层 ground floor

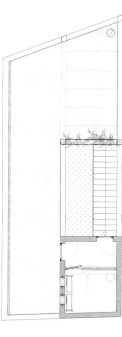

露台层 terrace floor

A-A' 剖面图 section A-A'   B-B' 剖面图 section B-B'   C-C' 剖面图 section C-C'

D-D' 剖面图 section D-D'   E-E' 剖面图 section E-E'

in those of sport, culture, social activities, and innovation in the fields of mobility and telecommunications.
Archea Associati has therefore opened a new studio in the residential district in Morumbi, between 9 and 15 km away from São Paulo's downtown. This project was Archea's first Brazilian architectural project.
This 3-storey building stands out from the neighborhood in development, because of its color : a dark, thick shade of brown blends with distinctive texture of the concrete. The palette-like graffiti artwork facades are placed directly on the sidewalk, opening a narrow road. On the other side of the building, the window, the entrance, the walls and the steel fence on the top are simply arranged.
The entrance leads to the shady lobby, where the large meeting room covered with corten steel is located.

项目名称：Archea Studio in São Paulo / 地点：Sumarezinho, São Paulo, Brazil / 建筑师：Archea Associati / 项目建筑师：Marco Casamonti
合作者：Luca Sartori / 客户：Archea Brasil / 有效楼层面积：100m² / 施工时间：2014.1—2015.3
摄影师：©Leonardo Finotti

114

Coldefy & Associés Architectes Urbanistes

Thomas Coldefy[right] and Isabel Van Haute[left] are leading the diverse international team based in Lille, France. Thomas studied in École Spéciale d'Architecture in Paris and Isabel in Saint-Luc in Ghent. They started their journey in famous French and American firms, DPA, SCAU, KPF and Richard Meier, before settling their practice in Lille. Is characterized by their dynamism and creativity, animated by an international tropism that drives them to participate regularly in professional events around the world as well as in large international competitions. Its international activity extends beyond Asia, having won the international conference center in Ouagadougou, Burkina-Faso in 2009, the Public Service Hall in Kobuleti, Georgia in 2012. Was awarded the silver prize for the IDA awards 2012. Since 2014, Thomas Coldefy is a visiting professor at the Institute of Architecture and Civil Engineering of Jilin in China.

108

Macro Sea

Created in 2009, Macro Sea is a design and development company based in New York City. Conceptualizes and builds creative interim use projects that transform and energize its surroundings. Its design projects have been featured in MoMA and the Venice Biennale. Builds several projects in New York City, including New Lab at the Brooklyn Navy Yard. Recently completed a campus that showcases a new model for study abroad in Berlin, and are researching new opportunities around the world.

162

Davidson Rafailidis

Was founded by Georg Rafailidis[left] and Stephanie Davidson[right] in 2009.

Georg Rafailidis studied at the University of Applied Sciences, Munich and AA School of Architecture, London. Has worked for the Herzog & de Meuron, London for 3 years before Davidson Rafailidis. Stephanie Davidson studied at the Mount Allison University, Dalhousie University and Architectural Association Diploma School. Has worked for the Sauerbruch Hutton Architects before Davidson Rafailidis. They lectured at the RWTH Aachen University and currently teaching at the SUNY Buffalo.

34
Torafu Architects
Was founded in 2004 by Koichi Suzuno[right] and Shinya Kamuro[left]. Koichi Suzuno was born in Kanagawa, 1973. Graduated from the Science University of Tokyo and completed the Master Course of Architecture at the Yokohama National University. Has worked at the Coelacanth K&H and Kerstin Thompson Architects. Is currently guest professor at the Kyoto Seika University and Ritsumeikan University. Shinya Kamuro was born in Shimane, 1974. Graduated from the Meiji University and completed the Master Course of Architecture at the same school. Has worked at the Jun Aoki & Associates for 4 years. Has been lecturer at the Showa Women's University since 2008.

22
debartolo architects
Principal and design leader, Jack DeBartolo graduated from the University of Arizona and Massachusetts Institute of Technology. Has worked with innovative desert architect William Bruder, supporting his firm for two years. In 1996, he joined his father, together forming the studio of Debartolo Architects, where they collaborated for 12 years. Is also teaching graduate students as a Faculty Associate of the Design School at ASU (Arizona State University) Herberger Institute for Design and the Arts. Debartolo Architects was named for the 2012 AIA Arizona Firm of the Year and Jack DeBartolo received the 2013 AIA Arizona Architects Medal.

150
FT Architects
Is a Tokyo based architectural office founded by Katsuya Fukushima and Hiroko Tominaga in 2003. They both graduated from the Tokyo National University of Fine Arts and Music in 1993. Katsuya worked for Toyo Ito & Associates and Hiroko worked for Hisao Kohyama Atelier prior to founding their office. They won the AIJ Award in 2015 with 'Timber Structure I and II" and taught at number of Japanese Universities. Currently Katsuya is an Assistant Professor at Tokyo City University and Hiroko is an Associate Professor at Kogakuin University.

56
CUAC Arquitectura
Was formed in 2006 by Javier Castellano Pulido[right] and Tomás García Píriz[left]. Both were born in Granada, Spain and graduated from the E.T.S. Architecture of Granada[ETSAG]. Tomás García Píriz started his own studio, 4:33 in Granada after graduation. Finished his postgraduate course at ETSAG in 2006 and received PhD degree in 2016. Javier Castellano Pulido started his own studio, Mesones 57 with Rubens Cortés, in Granada, 2002. Finished his postgraduate course at ETSAG in 2005 and received PhD degree in 2015. They have taught at the TU Berlin, Academy of Architecture of Mendrisio, EPFL (Switzerland) and ETSA Granada, Madrid, Valencia, Alicante, Sevilla.

66
Livil Architects
Liu Chongxiao gained master's degree of Architecture Design and Research from the Tianjin University. Is now living and working in Beijing. Has taught at the CAFA (Central Academy of Fine Art) as a Visiting scholar in 2005-2007 and worked at the CADG (China Architecture Design Group) as an architect in 2006-2015. Founded Livil Architects Design Studio in 2013, researches on material technology and structure based on mechanics, and hope to create the space in different scales of expression of modernity of China traditional artistic spirit.

76
ACDF Architecture
Senior associate architect, Maxime-Alexis Frappier graduated in 2000 from the School of Architecture, University of Montréal, winning the prestigious Canadian Architect Student Award of Excellence for his thesis work. Founded ACDF Architecture in 2006 and restructured in 2013. Now it includes four co-directors, Joan Renaud, Etienne Laplante Courchesne, Denis Lavigne and Yolande Jeanson. Senior architect, Joan Renaud graduated from the School of Architecture, University of Montréal and completed his Master Degree (CAO) in 2002. After then, he worked for the Saucier+Perrotte Architectes and for MSDL architectes. He joined ACDF in 2007.

Maxime-Alexis Frappier, Joan Renaud, Laure Giordani, Laurence Le Beux, Christelle Montreuil Jean-Pois, from left.

194

Archea Associati
Was founded in Florence, 1988 by Laura Andreini, Marco Casamonti and Giovanni Polazzi. They were born in Florence and graduated from the Faculty of Architecture of Florence. Silvia Fabi joined the firm in 1999. As well as headquarter in Florence, they have branch offices in 5 different cities including Milan, Rome, Beijing, Dubai and São Paolo. Besides the main activities of the office which consist of research on design and implementation of architecture at various scales, they reconcile intensive occupations of teachers and researchers in several universities in Italy within the area of architectural design.

42

Woods Bagot
Is a global design and consulting practice with a diverse portfolio of excellence spanning almost 150 years. The firm has a wide-ranging portfolio, with award-winning and globally-significant projects including the National Australia Bank headquarters in Melbourne, the International Renewable Energy Agency HQ in Abu Dhabi, and the Christchurch Convention Centre in New Zealand. A principal of global architectural firm Woods Bagot, Bruno's approach to architecture applies a comprehensive knowledge base spanning from conceptual design through to project delivery across a broad range of sectors, particularly education and lifestyle. He ensures continuous dedication to knowledge, research and the challenge of design methodologies is channeled via a commitment to lecturing and tutoring at local universities.

186

Hugh Strange Architects
Is a recently formed, award-winning practice based in London. Prior to establishing his practice, Hugh Strange graduated from the Edinburgh University in 1994 and worked for practices in London, Vienna and Berlin. The Strange House, the practice's first completed new-build project, won a RIBA Awards 2011, AIA UK Award 2011 (Best Small Project) and the Wood Award 2011 (Best Small Project). Was nominated for the 2015 European Prize for Contemporary Architecture - Mies van der Rohe Award and won RIBA National and Regional Awards. Is currently collaborating with the renowned Portuguese architect Alvaro Siza on a project adjacent to the Architecture Archive.

3XN
Danish Architect, Kim H. Nielsen[above] is Founding Partner, Principal and Creative Director of 3XN since 1986. Graduated from the Aarhus School of Architecture in 1981. Member of the Danish Architectural Association, RIBA and AIA. Jury Member of the World Architecture Festival and AR Awards for Emerging Architecture. Received numerous awards including RIBA Award 2005, 2007, 2009, 2011 and 2013. Is also Honorary Professor of Aarhus School of Architecture and Chairman of architecture committee at Danish Arts Foundation. Involved in all the major projects of the practice, including the Copenhagen Arena, Blue Planet Aquarium, Museum of Liverpool, Ørestad College and the UN City HQ in Copenhagen.

OMA
Shohei Shigematsu above is Partner at OMA(Office for Metropolitan Architecture) and the Director of the New York office. Was born in 1973, Japan and graduated from the Department of Architecture and received M.Arch at the Division of Engineering, Graduate School of Kyushu University. After graduation, he has worked at the Toyo Ito Architects & Assosiats and NKS Architects. Studied at the Berlage Institute in Amsterdam before joining OMA in 1998. Is a design critic at the Harvard Graduate School of Design, where recently conducted a research studio entitled Alimentary Design, investigating the intersection of food, architecture and urbanism.

Andreas Marx
Andreas Marx's expertise is based on social sciences and urban issues. A first degree in Sociology and Political Science at LMU Munich(2011) strengthened his knowledge of social interaction and the insights of social order. Following a part-time job as researcher for the Institute for Social Science Research, a strong interest for cities emerged and he continued studying at the UVA (Amsterdam 2012-13) - master degree in Urban Sociology. This led to a better comprehension of urban issues in modern cities. His Master Thesis deals with the perception of space and place - in detail with the occurrence and logic of a central daily food market in Munich. His academic focus lies on the question, how one may use the social sciences skills and architecture to improve future city development. Though he recently undertakes a Master degree in Urbanism - Landscape and City at TUM to learn more about the insights of architectural design and mapping in the field of urban planning.

Andrew Tang
Received his architecture degree at Institute of Design(IIT), Chicago in 1996. In the same year, he won the Jerrold Loebl Prize. Throughout his 20-year career, he has worked around the world contributing to many innovative and challenging projects in the field of Architecture & Urban Design. He has worked in the Netherlands on projects such as the new Central Station in Rotterdam serving as Architect, Urban Designer and Public Space Designer. Is currently a designer and founder of the design practice Tanglobe in Hawaii.

Javier de las Heras Solé
Graduated from the School of Architecture of Vallés in 2000 and College of Architects of Catalonia in 2002. Teamed up with the architects Jordi Bosch and Joan Tarrús in 2004. Has been awarded in numerous public competitions among which the new headquarters of Arquia, Bank of architecture in Girona. Is co-editor of the magazine Engawa which has been awarded in the AJAC VIII 2012 in the category "Promotion and dissemination of architecture". It was also invited to participate in forums in schools of architecture in Barcelona and Granada.

Christian Tonko

Was born in 1984 in Feldkirch, Austria. Studied architecture at the Academy of Fine Arts Vienna and Philosophy at the University of Vienna. Received B.A. from the University of Vienna and M.Arch. from the Academy of fine Arts Vienna in 2010. Has worked at the Wolfgang Tschapeller ZT Gmbh in 2010-2015 and founded his studio in 2016. Is currently writing his Ph.D. at the Institute for Art and Architecture, Academy of fine Arts Vienna. Now he lives and works in Vienna and Vorarlberg.

Izaskun Chinchilla Architects

Izaskun Chinchilla graduated in 2001 from the ETSA, UPM and has been driving her own office since then. She received Master level and Doctorte degree with a Magna Cum Laude in 2016 at the same university. She is Senior Teaching Fellow, Senior Research Associate and Public Engagement Fellow at Barlett School of Architecture. She has taught at the ESA Paris, Geneva University of Art and Design (HEAD), EPS Alicante, ETSA of UPM, CEU San Pablo University and IE University Segovia. Her research dedication has taken her as visiting scholar to Columbia University and Princeton University.

MAT Office

Is based in Rotterdam and Beijing, founded by Kangshuo Tang 唐康硕 and Miao Zhang 张淼 in 2013. Kangshuo Tang got the M.Arch from Harbin Institute of Technology in 2007 and graduated from Berlage Institute in 2013. Has worked at the URBANUS(Beijing) and NL Architects(Amsterdam). He is engaged as External Master Advisor by Graduate School and the School of Architecture, Harbin Institute of Technology. Miao Zhang graduated from Berlage Institute in 2013. Previously worked at URBANUS from 2005 to 2011. Is currently working and living in Beijing, doing various research and design projects, as well as teaching programs in optional studios.

© 2017大连理工大学出版社

版权所有·侵权必究

图书在版编目(CIP)数据

工作空间进化录：英汉对照 / 荷兰大都会建筑事务所等编；霍兴花译. — 大连：大连理工大学出版社，2017.7
 ISBN 978-7-5685-0912-1

Ⅰ. ①工… Ⅱ. ①荷… ②霍… Ⅲ. ①办公室－室内装饰设计－英、汉 Ⅳ. ①TU243

中国版本图书馆CIP数据核字(2017)第147380号

出版发行：大连理工大学出版社
　　　　　（地址：大连市软件园路80号　邮编：116023）
印　　　刷：上海锦良印刷厂
幅面尺寸：225mm×300mm
印　　张：13.5
出版时间：2017年7月第1版
印刷时间：2017年7月第1次印刷
出 版 人：金英伟
统　　筹：房　磊
责任编辑：杨　丹
封面设计：王志峰
责任校对：周小红
书　　号：978-7-5685-0912-1
定　　价：258.00元

发　行：0411-84708842
传　真：0411-84701466
E-mail：12282980@qq.com
URL：http://dutp.dlut.edu.cn

本书如有印装质量问题，请与我社发行部联系更换。